AGRICULTURAL ADMINISTRATION IN INDIA

ABOUT THE AUTHOR

Dr. K. VIJAYARAGAVAN (*b.* 1952), a well known scholar in the field of extension education is presently Senior Scientist at the Division of Agricultural Extension, Indian Agricultural Research Institute, New Delhi. He obtained his Master's degree in Agricultural Extension from Tamil Nadu Agricultural University, Coimbatore and Ph.D. from Indian Agricultural Research Institute, New Delhi. He has more than 17 years of teaching, research and extension experience at University level. His field of specialisation is development administration and extension management. He also has a wide experience in organisation of seminars, training programmes, workshops and conferences related to agricultural development.

Dr. Vijayaragavan is a prolific writer. Apart from his well-known book *Transfer of Technology to Small Farmers* (Concept, 1991), he has contributed more than 60 papers to various journals, national and international seminars, workshops and conferences. He is also an Editor of *Indian Journal of Extension Education.*

AGRICULTURAL ADMINISTRATION IN INDIA

A Comparative Study

K. VIJAYARAGAVAN
Senior Scientist, Division of Agricultural Extension
I A R I, Pusa Campus
NEW DELHI — 110012

CONCEPT PUBLISHING COMPANY, NEW DELHI 110059

The Indian Council of Social Science Research gave the financial support towards the publication of this work. However, the responsibility for the contents of the book — facts, opinions, conclusions, etc. — is entirely that of the author and not of the ICSSR.

ICCSR Consultant: **Dr. G. Perumal**

ISBN 81-7022-519-1

First Published 1994

Published and Printed by
Ashok Kumar Mittal
Concept Publishing Company
A/15-16, Commercial Block, Mohan Garden
NEW DELHI-110059 (India)

Lasertypeset by
Printing Express
H-13, Bali Nagar
NEW DELHI 110015

ACKNOWLEDGEMENTS

I owe gratitude for the valuable help, support and encouragements in various forms to so many people that the placing on record of my appreciation to all of them seems impossible. Although I can only mention a few, I wish to thank all those unmentioned who in different ways have contributed to my work and helped me to complete my task. Dr. Y.P. Singh, Professor, Division of Agricultural Extension, Indian Agricultural Research Institute, New Delhi, Chairman of my Advisory Committee has been a source of great inspiration to me through his valuable guidance, suggestions, ungrudging help and affectionate behaviour throughout the period of the study. For many years I have been fortunate to enjoy a close and fruitful association with him who has not only stimulated my interest in the area of development administration but has also given me constant support.

I am highly indebted to Dr. B.P. Sinha, Head, Division of Agricultural Extension and member of my Advisory Committee for his constant encouragement, help and suggestions throughout the course of investigation. Dr. Mruthyanjaya and Dr. O.P. Kathuria, members of my Advisory Committee, have taken great pains in rendering valuable advice, suggestions and comments, I am highly thankful to them.

I was given a very good co-operation for the conduct of the study in both the states — Madhya Pradesh and Tamil Nadu. I am particularly grateful to Shri N.N. Mehta and Shri N. Athimoolam, IAS, Directors of the department of agriculture in the above states for granting permission to conduct the study and making provisions for the collection of data. I wish to express my sincere thanks to all the officials of department of agriculture of Madhya Pradesh and Tamil Nadu who have participated or helped in this research endeavour without whose active co-operation this study could not have been materialised.

The kind help of Shri A.K. Dubey in translating the interview schedule into Hindi is appreciated. Also the encouraging words and timely help of all the scientists and staff of the Division of Agricultural Extension, particularly Dr. R.P. Singh, Dr. A.P. Singh, Dr. A. Ramesh Babu, Dr. O.P. Kataria, Shri T.P. Singh and Mrs. Rekha Bhagat can not be forgotten. I owe a lot for the concern and help of my brothers Joy Thomas, Sam Russel, Lakshman Kumar and K.P. Kannan. The generosity and goodness of Dr. Clifford Kumar, Mr. Prabhakaran and Mr. C. Muthukani and his family were an immense source of help for my data collection work at Madras and Bhopal and I thank them.

I am grateful to my beloved parents for their love, affection, guidance and encouragement which gave my strength to complete the study.

I thankfully acknowledge the publication grant awarded by ICSSR, New Delhi without which the publication of my work could not have been materialised.

While acknowledging my appreciation for those in the Concept family, I forget not to mention Shri Ashok Kumar Mittal who took special interest in bringing out this publication.

My wife Lalitha, deserves a very high appreciation for her untiring help and active co-operation in a number of ways which can not possibly be expressed in words. So, thank you Lalitha, for your unfailing loyalty and love.

K. VIJAYARAGAVAN

CONTENTS

LIST OF TABLES

LIST OF FIGURES

ABBREVIATIONS USED

AAO	=	Assistant Agriculture Officer
AO	=	Agricultural Officer
ADA	=	Additional Director of Agriculture
ADO	=	Agricultural Development Officer
DA	=	Director of Agriculture
DDA	=	Deputy Director of Agriculture
DS	=	Developed State
FDO	=	Field Demonstration Officer
JDA	=	Joint Director of Agriculture
JLAO	=	Junior Level Agricultural Officer
LDS	=	Less Developed State
MP	=	Madhya Pradesh
RAEO	=	Rural Agricultural Extension Officer
SADO	=	Senior Agricultural Development Officer
SDAO	=	Sub-Divisional Agricultural Officer
SLAOs	=	Senior Level Agricultural Officers
TN	=	Tamil Nadu
T & V	=	Training and Visit of system of Extension
VLW	=	Village Level Worker

1

INTRODUCTION

One of the most important factors of development of agriculture in countries like India is the effective administrative organisation which can take the benefits of technological advancements to the doorsteps of farmers. Realising the importance of administrative organisation, the Report of the Agricultural Administration Committee (1958) observed that "the problem of increasing the agricultural production in India is more an administrative problem than a technical one". Inefficient administration has greatly contributed towards shortfalls in implementation of agricultural schemes and has resulted in lower production. An efficient agricultural administrative system is needed to improve the overall agricultural production and productivity.

Various committees and commissions have studied the problems of agricultural administration of India and emphasised the importance of efficient agricultural administration. For example the Joint Indo-American Team on Agricultural Research and Education (1955) observed that some of the major difficulties in the way of developing improved technical leadership in the fields of agricultural research and advisory services were due to poor policies on administration and personnel management. The Report on India's Food Crisis and Steps to Meet it (1956) has also pointed out the importance of effective co-ordination in agricultural administration as well as the need of powerful and energetic organisation in increasing food production. It stated that extraordinary organisational and administrative measures and actions are required to help the nation to meet the challenge of producing enough food to meet minimum requirements of a rapidly growing population.

The problem of co-ordination in agricultural administration was discussed at the Annual Conference of State Ministers of Community Development and Panchayati Raj held in 1963. This led to the setting up of a working group with Shri Ram Subhag Singh, then Union Minister of Agriculture as Chairman to study the problems of administration and co-ordination in agriculture. The working group recommended the formation of Agricultural Production Committees at State, District and Block levels, each headed by an appropriate official to co-ordinate the activities of all relevant departments at that level.

During 1967, Administrative Reforms Commission of India set up a study team on agricultural administration to go into the various problems related to agricultural administration. The study team was headed by Shri H.M. Channabasappa. It has made a number of recommendations in its Report on Agricultural Administration (1967) about the organisation and administration of agricultural research, education, extension, finance, supplies and services.

A review of the history and growth of agricultural department at the centre and at state will reveal the progress and problems experienced by India in the organisation of agricultural development. The realisation that the formation of an effective government machinery for administering agricultural development is essential, was felt in India as early as 1866 as a result of report of the Famine Commission as well as the recommendations of Manchester Cotton Association during 1869. As a result, a Department of Revenue, Agriculture and Commerce was established in 1871 at the Centre as a branch of the Secretariat of Government of India. During 1879 for reasons of financial stringency, the Department was merged with the Home Department. However, during 1881 a separate Department of Revenue and Agriculture was revived in the Central Secretariat as a result of the recommendations of the Famine Commission of 1880-81.

Realising the importance of Provincial Department of Agriculture, Royal Commission on Agriculture (1928) observed that "without provincial agency no programme of development which emanated from headquarters could be productive of tangible results". Provincial Departments of Agriculture were established in many Provinces during the period between 1875-1905 following the report of Famine Enquiry Commission of 1880. The growth was parallel to that of the Central Department of Agriculture. The functions of the Department of Agriculture were mainly concerned with statistical enquiries, land records and famine relief. But following the report of Dr. J.A. Voelcker on 'Improvement of Indian Agriculture' the activities of the Central and Provincial Departments were strengthened in the areas of agricultural research and education. The scientific and developmental staff of the Provincial Department was further strengthened following the report of Famine Commission of 1901. During Lord Curzon's viceroyalty (1898-1905) greater financial allocation was made to agricultural departments and also full time Directors of Agriculture were appointed in major Provinces. The Provinces were divided into suitable Circles and each Circle had a Deputy Director of Agriculture who was incharge of the activities of experimental work as well as demonstration or extension activities.

The Imperial Agricultural Service was constituted in 1906. However, the service was ceased in 1924 following the Reforms of 1919 when agriculture was transferred as a state subject.

During 1930 the departments of agriculture were eager to follow the recommendations of Royal Commission on Agriculture but were handicapped by lack of financial resources. During 1940, following the food crisis created by the Second World War and the Bengal Famine there was a rapid increase in the outlay for agricultural department. The extension or developmental machinery at the field level also got greater attention. After Independence, during the Plan periods the administrative apparatus of the agricultural departments in various states were strengthened and modified following the introduction of many programmes and technological advances in agriculture. As a result, they have grown in their size with increased budget and area of operation. Recently the introduction of Training and Visit System of extension in most of the states in India has also resulted in a number of organisational changes in the department of agriculture which is aimed at increasing the overall efficiency of the department.

Agricultural departments have now been well established in all the states of India. But their structure, functioning and efficiency have been varying from state to state and generally found to be not satisfactory even from their very inception. Further their functioning has been hampered by many factors like lack of financial resources, lack of co-operation of farmers and unfavourable weather conditions. For example, the Report on the Improvement of Indian Agriculture (1897) made the following observation :

> The organisation of Provincial Departments of Agriculture is not alike throughout India. In the Punjab for instance there is no separate departments of agriculture, but it merely forms a part of the Land Revenue Administration, its Report being included in the general one of the Administration and not being given under the different head prescribed by the Government of India.

Highlighting the lack of participation of farmers in the activities of agricultural department in India the Report on the Improvements in Indian Agriculture (1909) made the following observation :

> The essential difference between the agricultural departments in the East and in the West is that the latter have arisen to meet the spontaneous demand of cultivators of the soil, where the former have lately been created by a bureaucratic Government anxious to give all assistance it can to its agricultural subjects. The demand for improved agriculture has not in India, except in special cases, come from the cultivators. While, therefore, in the West the cultivator is naturally in direct touch with the department of agriculture, in India, it is necessary for the department to put forth every effort first to ascertain the needs of the cultivators and then to demonstrate how they can most effectively be met.

The Royal Commission on Agriculture (1928) reported that the work of the department of agriculture was greatly hindered by three main causes namely want of sympathy, imperfect machinery and want of money.

The lopsided development of agricultural departments, especially during their early period by confining their activities mainly to agricultural research and few demonstrations was pointed out as early as 1937 by Sir John Russel. He observed in his Report on the work of the I.C.A.R in applying science to crop production in India (1937) that "in India a state has been reached where the machinery for gaining more knowledge is working better than the machinery for utilising it."

Apart from the problems of lopsided development, the problems of differences among the agricultural development with regard to its resources can be seen from their expenditure and staffing pattern. Even from the beginning they had wide differences among themselves in their pattern of expenditure, staffing and achievements. While some Provinces like Madras and Bombay were strong in agricultural research and education, Provinces like Orissa and Bihar were relatively weak in agricultural research. During 1926-27 the departments of agriculture of Madras and Bombay spent about Rs. 16.5 lakh each, whereas the combined expenditure of departments of agriculture of Bihar and Orissa was only Rs. 6 lakhs. During 1937-38, the departments of agriculture in Madras and Bombay had a total technical staff strength of 398 and 314 officers respectively. But the technical staff strength of the agricultural departments in the Provinces of Bihar and Orissa was only 136 and 13 officers respectively.

The differences in the organisation of agricultural department, its resources as well as overall agricultural development between the states have widened after Independence. The agricultural development of the states also depend upon a number of factors like availability of irrigation, technologies and inputs. Further the problems of agricultural administration contribute to the lower agricultural production and development in the states. With all the problems of agricultural development, some states like Punjab have managed to come up better in agriculture than other states. The reasons for the agricultural development in these states could be attributed to an efficient agricultural administration apart from other factors like availability of irrigation, infrastructural facilities and other resources. Though various committees and a few studies have pointed out the general problems of agricultural development administration none has studied and compared the structure, processes, functioning and efficiency of agricultural administration of agriculturally developed and less developed states. Does the agricultural administration of the developed state significantly differ from less developed state? If yes, in what dimensions? There is great need to find answer to the above question which would not only give a clear picture of the agricultural development administration in the developed and less developed state, but also become a basis for taking effective steps to increase the overall efficiency of agricultural development administration especially in the less developed states. Therefore, the present study has been undertaken to answer the above questions. The specific objectives of the study are as follows :

1. To develop agricultural development index of various states.
2. To study the selected dimensions of organisational context, and macro-organisational structural configuration of state department of agriculture in the developed state and the less developed state.
3. To study the personnel management practices like recruitment, promotion and salary of the employees of the department of agriculture in the developed and the less developed state.
4. To study the selected dimensions of unit structural characteristic, intra-unit communication, workgroup functioning, job technology, supervisory behaviour, pay administration and creativity climate of state department of agriculture in the developed state and the less developed state.
5. To study the overall organisational effectiveness of agricultural department of the developed state and the less developed state.
6. To find out whether the employees of the department of agriculture of the developed state and the less developed state

can be distinguished on the basis of selected variables.

Scope and Importance of the Study

The study would answer the question whether the structure and processes of agricultural development administration in the developed state and the less developed state differ from each other or not? If yes, on what dimensions? The insight into the differentials of agricultural development administration between the developed and the less developed state would be a basis for making necessary changes or improvement in the agricultural development administration of the less developed states. Further the agricultural development index of various states would point out the comparative position of various states with regard to agricultural development. This would be a pointer towards making greater efforts on investment towards agricultural development especially in less developed states. The present investigation is one of the few studies in the area of agricultural development administration which has gone in depth to study the structure, processes and effectiveness of agricultural departments.

Limitations of the Study

The present investigation suffered from the general limitations faced by any Ph.D. scholar in terms of availability of time, funds and other resources. The study was conducted in two states (one developed and another less developed state), and it would be better to conduct such study in more states to make generalisations which will be applicable to larger areas.

Organisation of the Report

The entire report of the investigation has been presented in eleven chapters. The findings have been presented according to the objectives of the study, each presented under a separate chapter. The findings and discussion have been merged together to enhance the comprehension and communicability. The important contents of each chapter are given below :

Chapter I : It highlights the importance and relevance of the present investigation and narrates the specific objectives of the study.

Chapter II : This chapter gives the theoretical orientation of the study and deals with review of literature relevant to the investigation. It also presents the hypotheses of the study.

Chapter III : This chapter describes the research methodology including the measurement of various variables followed for the investigation. The statistical methods used in the study have also been given.

Chapter IV : This chapter gives an account of development of an Agricultural Development Index of various states of India which formed a basis for selection of an agriculturally developed state and a less developed state for the present study.

Chapter V : The organisational context of the department of agriculture of the developed state and the less developed state has been described in this chapter.

Chapter VI : It discusses the macro-organisational structural configuration of the department of agriculture of the developed state and the less developed state.

Chapter VII : The personnel management practices like recruitment, salary and promotional policies of the department of agriculture of the developed state and the less developed state have been described in this chapter.

Chapter VIII : This chapter gives an account of the selected dimensions of unit structure and processes of the department of agriculture of the developed state and the less developed state.

Chapter IX : The results of the study of the organisational effectiveness of the department of agriculture of the developed state and the less developed state have been presented in this chapter.

Chapter X : This chapter gives the results of discriminant function analysis employed to find out whether the employees of the department of agriculture of the developed state and the less developed state could be discriminated on the basis of selected variables.

Chapter XI : This last chapter gives the summary of the findings of the study and their implications. It also gives suggestions for future research undertakings.

2

THEORETICAL ORIENTATION

The theoretical or conceptual perspective of various variables studied in the present investigation has been discussed in this chapter. Along with the conceptual perspective, a brief review of relevant literature pertaining to the study has also been done. It should be mentioned here that only few works have been conducted/reported on the comparative study of development administration in the developed and less developed regions.

The studies of Mathur (1982)[1] and Pai Panandikar and Kshirsagar (1978)[2] showed that some aspects of bureaucratic structure and processes differed among regions of different development levels. The Working Group on organisational structures set up under National Committee on Development of Backward Areas[3] requested the Kerala State Planning Board to undertake a study on development administration of a comparatively backward District and a forward one. The study undertaken in Malappuram (backward District) and Eranakuam (forward District), had the following objectives:

(a) To study the availability of technical personnel of various qualifications and their equation with development programmes.

(b) To investigate whether by a proper reallocation of duties and locations both of individual and departments it might be possible to get more work done.

(c) To analyse strength, qualifications and job content of various categories of job in the development department. The major findings of the study were as follows:

(i) In both the Districts the multiplicity of agencies with overlapping functions was found. This was due to the trend of over-departmentalisation and as a result certain programmes were concurrently implemented by several departments;

(ii) The total number of field units under Deputy Director of Agriculture was higher in forward District than that of backward District. However, interestingly the total number of qualified personnel was high (84) in backward District than that of forward District (51);

(iii) The number of posts which were vacant was high in backward District than forward District. For example, as many as 60 out of 117 posts of General Extension Officers and 49 out of 69 posts of Cattle Improvement Assistants were laying vacant in backward District;

(iv) The effective co-ordination among various agencies of agricultural development was lacking in both the Districts; and

(v) District Planning was found to be done without the participation of field level functionaries.

The conceptual framework of the study has been discussed below:

Organisational Context

Conventionally organisational context refers to all conditions and factors external to organisation or unit under consideration (Blau and Schoenherr, 1971). In this sense context is closely related to earlier definitions of environment as the set of constraining phenomena that exit external to organisation but within which the organisation must function (Van de ven *et al.*, 1974). Expanding upon the definition of Pugh *et al.* (1969), Ven de ven and Ferry (1980) defined organisational context as 'social and economic setting in which an organisation chooses to operate. In other words context is defined as those dimensions within the control of an organisation that create the setting for the development of organisational structure'. The context of any organisation include domain factors like organisation age, history, domain type (types of functions performed, products or service rendered) and economic factors like demand for products or service, number of clients, size of resources (budget and production capacity). For the present investigation the following two contextual factors were studied with respect to agricultural department in the developed state as well as the less developed state:

1. Origin and history of the department
2. Budgeted expenditure

Macro-Organisation Structure

The term 'structure' refers to the pattern in which various parts or components of a whole is arranged. Organisational structure can be defined as the network of durable and formally sanctioned organisational arrangement and relationship (Miller, 1965). The organisational structure being abstract is not visible in the same way as a biological or mechanical structure. But it can be seen from the actual operations and behaviour of the organisation. Organisational structure performs the following three major functions (Khandwalla, 1977) :

(i) If affords the organisation the mechanisms with which to reduce external and internal uncertainty;
(ii) It enables the organisation to undertake a wide variety of activities through devices such as departmentalisation, specialisation and delegation of powers; and
(iii) It enables the organisation to keep its activities co-ordinated, to pursue goals and to have a focus in the midst of diversity.

At macro-level the structure is defined in terms of skeletal division of labour and authority in an organisation illustrated by the organisational chart. Its underlying dimensions are size, degree of vertical and horizontal and spatial differentiation. The macro-organisational structural dimensions investigated in the present investigation are given below:

Organisational size: It refers to the number of full time salaried employees in an organisation (Blau and Schoenherr, 1971; Hikson *et al.*, 1971; Hall *et al.*, 1987). The increase in organisational size results in increased division of labour (tasks and functions to be performed), available resources (personnel, money and technologies) and allocation of employees to different organisational units. The division of labour makes specialisation possible and enables an organisation to overcome the limited set of skills of any one individual (March and Simon, 1958). However, the benefits obtained from specialisation that comes with the division of labour are limited particularly for labour intensive organisations. Van de ven and Ferry (1980) in their study on government departments found that office size was slightly positively associated with efficiency and effectiveness but it was significantly negatively correlated with job satisfaction of office personnel. A mixed and non-significant relationship have been found among many studies which directly related organisational size to various measures of organisational productivity or goal attainment (Berger and Cummings, 1979).

Number of supervisory level: Major indicators of vertical differentiation is the number of hierarchical levels. The concept of supervisory or hierarchical levels derives from the principles stated by Weber (1947) : "The principles of office hierarchy and of levels of graded authority mean a firmly ordered system of super and sub-ordination in which there is a supervision of the lower offices by the higher ones". It is also called scalar chain principles (Fayol, 1949). The number of supervisory level is measured by counting the longest chain of command in organisation chart (Van de ven and Ferry, 1980).

Number of job titles: Number of job titles is one of the indicators of horizontal differentiation. It indicates the overall division of labour among formal positions in an organisation. The number of different service classifications not counting different grades within a classification is used as the measure of number of job titles in the organisation (Van de ven and Ferry, 1980). It has been reported that decrease in number of hierarchical level (or vertical differentiation) and increase in horizontal division of labour result in increase in the performance of government department.

Number of geographical operating units: It is the indicator of spatial differentiation of the organisation. It is measured by counting the total number of geographical operating units of the organisation. Increase in geographical division of labour generally leads to increased administrative overhead costs in each operating unit. But spatial differentiation increases the accessibility and adaptability of the organisation to its clients and suppliers. Increasing the number of geographical service units will be less costly for labour intensive organisations than capital intensive organisations. It will also result in increased decentralisation of powers and supervisory intensity at field level.

Van de ven and Ferry (1980) reported that overall performance of government departments was found to have strong association with the increase in horizontal division of labour (particularly more geographical sites).

Supervisor-staff ratio: One of the indicators of administrative intensity is the supervisory-staff ratio computed as the total number of supervisors divided by the total number of non-supervisory personnel in an organisation.

Percentage of supervisors above bottom level unit supervisors: It is a supplementary indicator of administrative intensity which is the percentage of all supervisors in an organisation who are located hierarchically above bottom level unit supervisors. This indicator reflects the proportion of supervisors who are middle and upper level managers and who are not directly supervising and at production level or field level (Van de ven and Ferry, 1980). The proportion of middle and upper level supervisors is obtained by dividing the total number of supervisory positions into the number of supervisors above the first line unit supervisors. Studies have pointed out that overall performance of government departments was found to be strongly associated with decrease in administrative intensity or overhead at middle and upper level of hierarchy (this is obtained by decreasing the percentage of supervisors in position above bottom level unit supervisors) and by increase in administrative intensity and supervision at lower technical level of field units (this is obtained by increasing the supervisor-staff ratio).

Technical employees-staff ratio: It indicates the intensity of technical employees in the organisation and refers to the ratio between the technical employees and the total non-technical employees of any organisation.

Ministerial employees-staff ratio: It indicates the intensity of ministerial (administrative) employees in the organisation and refers to the ratio between the ministerial employees and the non-ministerial employees of the organisation.

Supporting employees-staff ratio: It indicates the intensity of supporting staff (class IV employees of the state department) and refers to the ratio between supporting employees and the total non-supporting employees of the department of agriculture.

Percentage of posts filled into the total number of posts sanctioned: It indicates the percentage of technical posts filled into the total number of technical posts sanctioned.

Personnel management practices: Personnel management according to the definition adopted by the Indian Institute of Personnel Management

(1962) is "that part of management function which is primarily concerned with the human relationships within an organisation. Its objective is the maintenance of those relationships on a basis which, by consideration of the well-being of the individual, enables all those who engaged in the undertaking to make their maximum personal contribution to the effective working of the undertaking". The area of personnel management encompasses methods of recruitment, selection, training, wage and salary administration, employee assessment and promotion, employee service conditions, employee record and labour relations (Famularo, 1972; and Agarwal, 1973).

The vitality of any organisation depends heavily upon the quality of its personnel management. Personnel is so important and so sensitive amongst all the factors of production that it demands utmost attention and consideration. The motivation and performance of the employees depends upon the personnel management practices of any organisation. In the present investigation personnel management practices in the department of agriculture with regard to recruitment, salary and promotion were studied.

Unit Structural Characteristics and Processes Unit Structure

Most of the approaches to organisational structure have been based on Weber's (1947) idea of bureaucracy or on ideas about management of organisations written by writers like Gulick and Urwick (1937) and Fayol (1949). In order to design reliable measures of different aspects of organisational structure, Pugh *et al.*, (1968) at the University of Aston combined the above approaches and constructed different scales to measure the organisational structure. Their scale measured the following five primary dimensions of organisational structure : specialisation, formalisation, standardisation, centralisation and configuration. Most of the studies using the above scales in measuring the structure is done based on composite scores across the organisation. But as we know a complex organisation consists of many sub-systems. Each sub-system has a structure corresponding to its tasks and functioning and contributes to the total system. Thus measuring organisational structure by composite scores based on dimensions such as specialisation, formalisation and discretion across all organisational units present a distorted homogenizing view of organisation (Van de ven and Ferry, 1980). This calls for measuring organisational structure at unit level.

The organisational unit can be defined as an unit consisting of a superior and all personnel reporting directly to that supervisor. Organisational unit bears many resemblances to large scale social system. A review of literature will show that the basic structural referent of organisational unit is its work programme and its dimensions are examined in terms of specialisation, standardisation, discretion and expertise (Perrow, 1967; Hall, 1962; Hage, 1965; Van de ven, 1976). The following unit structural characteristics have been examined in the present investigation:

Unit standardisation: It refers to the degree to which work rules, policies and procedures are formalised and followed in an organisational unit. Standards are formalised when they are codified and specified in detail. Standards are followed when unit personnel perceive that use of formal or informal operating practices in their work (Hall, 1962; Hage and Aiken, 1969).

Distribution of unit authority: It refers to the broader perspective of centralisation. Centralisation refers to the locus of decision-making authority in an organisation. When most of the decisions are made hierarchically, an organisation unit is considered to be centralised. Centralisation has been measured as the degree of hierarchy of authority on work related decisions (Hall, 1962; Hage and Aiken, 1969; Khandwalla, 1977). But it has been observed that a number of additional sources of decision-making authority exist within organisation. Max Weber (1947) has pointed out that in addition to bureaucratic hierarchy, authority can be exercised by 'collegial bodies'. The multiple sources of decision-making have been recognised by Duncan (1972) and Wood (1973). Thus the 'distribution of unit authority' in this present study refers to the distribution of authority among the unit supervisors (hierarchical authority), unit member (personal authority), and staff or line positions outside of the organisational unit (functional authority) (Van de ven and Ferry. 1980).

Role interchangeability of personal specialisation: Role interchangeability means the degree the which 'A' can perform B's job at short notice and 'B' can perform A's even when 'A' and 'B' have different job titles or functional assignment (Beer, 1966; Tyler, 1973). Job rotation will be difficult if there is higher personal specialisation in a unit. Thus personal specialisation refers to bifurcation of skills among unit employees (Blau and Schoenherr, 1971). It is important to note here that the level of expertise or professionalism is conceptually independent of personal specialisation.

Unit incentives: The unit incentives refers to the degree to which individual or group performance results is rewarded for doing work (Van de ven and Ferry, 1980).

Unit sanction: It refers to the extent to which individual or group will be punished for doing a poor job.

Design of Jobs

At micro-level, the job or position that individual occupies form the core focus of any indepth study of any organisation. The design of job or position has been the central concern of organisational researchers since the advent of scientific management theory (Taylor, 1911), administrative management theory (Fayol, 1949) and bureaucratic theory (Weber, 1947). The way in which the job is structured is important for the following reasons : (1) It is basic criterion for personnel selection, training, promotion and pay (e.g. Ghiselli and Brown, 1955; Guion, 1965), (2) It defines and controls the roles and behaviour of individuals who occupy the jobs or positions (Weber, 1947; Argyris, 1964; Emergy and Trist, 1960) and (3) It influences the organisational productivity, effectiveness, job satisfaction and general quality of working life (Trist, 1970; Hulin and Blood, 1968; Hackman and Lawler, 1971).

The classical management theories laid emphasis on simplifying, specialising and standardising jobs in order to increase work performance. But research by a number of scholars proved that simple, routine and non-challenging jobs would lead to employees' dissatisfaction and lower efficiency (Hulin and Blood, 1968). As a result many organisational experts began to argue for job expansion or enlargement (Trist, 1970; Argyris, 1957; Mausner and Snyderman, 1959). However, the theories which advocated job enrichment were criticised for the following two basic limitations:

(i) The theories assume that a normal person exist. They do not admit that individuals vary in the degrees to which they desire challenging, risky, responsible and accountable work settings (Porter *et al.*, 1975). Many recent researchers have confirmed that certain individual characteristics must be taken into account along with job characteristics to generate valid predictions about the employees' behaviour (Blood and Hulin, 1967; Hackman and Lawler, 1971; Oldham *et al.*, 1976). These individual characteristics may include person's age, sex, organisational tenure, family size and income; and

(ii) With the notable exception of few researchers like Trist *et al.* (1970) the experts have tended to design job in an organisational and work group vacuum.

Only recently attempts have been made to integrate and examine the connections between design of jobs, work groups and of the overall organisations (Pierce, 1978). As pointed out by Morse and Lorsch (1970) employees motivation and unit performance would be highest when there is congruency between the structure of the organisation and the structure of the employee's job. Recent organisational research on work design has succeeded in identifying a set of perceived job characteristic that holds meaning for job satisfaction and contributes to performance. Hackman and Oldham (1975) developed a Job Diagnostic Survey which measured the following job dimensions : (i) skill variety, (ii) task identity, (iii) task significance, (iv) autonomy and (v) feedback. Realising that the Hackman and Oldham job dimensions focused heavily on attitudinal aspects of job, Van de ven Ferry (1976) proposed the following dimensions which emphasised on behavioural aspects : (i) job specialisation, (ii) job expertise, (iii) job standardisation, (iv) job discretion and (v) job incentives. In the present study the following behavioural aspects on job design as suggested by Van de ven and Ferry (1980) have been studied with respect to the employees of the department of agriculture. It should be mentioned here that the job design dimensions studied here are analytical properties or characteristics of members which are defined and measured without reference to characteristics of the group.

Job expertise: It refers to the skill level of employees of the organisation in terms of formal education, length of job entry, orientation and training .

Job standardisation: It refers to the degree to which roles and tasks are spelled out in job description and the degree to which rules and procedures are defined. Job standardisation focuses on the clarity and details of guidelines regarding the duties and responsibilities of the position and methods for performing these duties.

Job authority: It refers to the amount of discretion or influence that the job incumbent exercises in making job related decisions.

Job pressure: It refers to the amount of work pressure involved in the job. It includes the individual's control over job deadlines, the degree to which the individual is constrained by present quality standards and the amount of lead time given on incoming work.

Job accountability: It is defined as the extent to which the job incumbent answers for or holds him/herself responsible for work decisions and job behaviour.

Job feedback: It refers to the amount of information about the procedures and results and employee receives from his/her job, from other employees or colleagues and from the supervisor.

Expectation of rewards: It refers to the extent an employee perceives that doing a good job will be rewarded.

Expectation of sanctions: It refers to the extent to which an employee perceives that doing a poor job will be punished.

Intra-unit Communication or Information Flows

Communication is one of the important processes of organisation. It is a complex process involving many dimensions (Hovland, Janis and Kelley, 1953; Hovland *et al.*, 1957). Chester Barnard (1938) assigned a central place to communication in organisational effectiveness. A number of other behavioural scientists have underlined the importance of communication in contributing to organisational effectiveness (Bavelas and Barret, 1951; Dorsey Jr., 1957). Communication is essential for internal functioning of the organisations. It is required to establish goal, develop plans, organise human and other resources, and develop and appraise members. Willits (1967) found that the high performing organisations have less guarded upward flow of ideas.

In the study by Indik, Georgopoulos and Seashore (1961) it was reported that higher degree of performance and effectiveness of non-supervisory employees were associated with higher degree of communication climate as perceived by the employees. Macdonald and Farance (1970) studied the frequency, initiation, content and perceived freedom of communication of employees with supervisors. The study pointed out that under unfavourable communication climate two way communication behaviour is restricted, information accuracy and adequacy becomes highly questionable and employees satisfaction and company performance is restricted. The concept of information flow refers to messages or communications about the materials or work objects that are transmitted between people through a variety of media. Transmission of information may occur through media like signs, rules, reports, personal discussion and group or committee meetings (Van de ven and Ferry, 1980).

Information flow is one of the basic elements of process in organisation. It is the living functionary of the organisation. The amounts of information flow indicate the intensity of intra-organisational network. The variability or standardisation of information flow indicates the routinisation of network relationship. In the present investigation the following information or communication flows of the department of agriculture were studied :

Written communication flows: It refers to the work related messages sent among employees of the organisation through written memos, reports and letters.

Personal communication flows: It refers to the work related messages sent among employees of the organisation through personal one-to-one discussions.

Group communication flows: It refers to the work related messages sent among employees of the organisation through group or staff meetings.

Work Group Functioning

Work groups are important features of work environment. It comprises of a set of people who share the same or adjoining work places, report to the same supervisors and engage in tasks that are similar or related (Cammann *et al.*, 1983). The importance of group phenomena in organisational functioning has been amply documented (Likert, 1961; Hackman and Morris, 1975). It has been found that group forces are important not only in influencing the behaviour of individual work groups but also in affecting the behaviour of entire organisation. Georgopoulos (1957) has demonstrated that group forces in the form of group norms are related to performance of industrial organisations.

The experimental study of Coch and French as reported by Likert (1961) confirms the power of group goal which can influence the productivity of the organisation. There are ample evidences to show that the productivity of the organisations can be increased when the goals of work groups are consistent with the objectives of the organisation (Likert, 1961). Katz *et al.* (1950) and Katz *et al.* (1951) found that work groups with greater loyalty and cohesiveness were producing at higher level. Lewin (1958) has reported that work group with high peer group loyalty and common goals are effective in achieving their goals. The analysis of Seashore (1954) provides impressive evidences of the power of the goals of cohesive groups. His study showed that greater the cohesiveness of group, the greater the influence which the goals of the group have on the performance of members of the groups. The group cohesiveness has also been found to contribute to the greater productivity and higher job satisfaction (Blau and Scott, 1962).

The above brief review of research on work group functioning will reveal that various dimensions of work group functioning like group goal, group loyalty and group cohesiveness, have been studied in various investigations. In the present investigation the following aspects of work group functioning as suggested by Cammann *et al.* (1983) were studied:

Group homogeneity: It is the degree to which members of the same group are similar in their skill, abilities and backgrounds.

Group goal clarity: It refers to the degree to which the members of the group are clear about the goal of the groups.

Group cohesiveness: It measures how attractive the group is to the members.

Open group process: It refers to the degree to which the members of the group freely share their feelings, express their view and involve in making group decisions.

Internal fragmentation: It refers to the degree of interpersonal friction exist among the group member.

Supervisory Behaviour

The work performance of any organisation is affected by the type of supervisory or leadership styles. The leadership is the process of influencing the activities of an individual or a group for goal achievement in a given situation. The supervisor influences the nature of tasks to be performed, helps to form goals, mediates rewards, control structures and serves as an agent in policy and information transmission (Cammann *et al.*, 1983).

Various studies have pointed out that supervisory behaviour of high production units is different from that of low production units. Kahn (1956) found that foremen incharge of high producing units are both employee centered and seen by them as feeling that achieving a high level of production was one of the most important parts of their job. Likert (1961) found that close supervision was more often associated with a high rather than a low level of productivity. Similar results have also been reported by Katz *et al.* (1950).

It has been reported by many studies that supervisors of high productive units focussed more upon human aspects of their sub-ordinates as compared to job-oriented functions. They also spent more time in motivating the employees, providing structure and training the sub-ordinates (Spotts, 1976; Bass, 1960; Likert and Kahn, 1956; Mann and Dent, 1954; Katz *et al.*, 1951). Mann and Baumgartel (1953) found that supervisors who made their men feel free to discuss job problems with them had less absence in their work group than did supervisors who displayed opposite behaviour. Different supervisory behaviour or managerial dimensions have been studied by different investigators. For example, Bowers and Seashore (1966) studied the dimensions like 'support', 'interaction facilitation', 'goal emphasis' and 'work facilitation'. Other dimensions studied are as follows : Consideration, sensitivity, production emphasis, initiation structure, employee orientation, production orientation, closeness of supervision, group relationship, enabling goal achievement, administrative skill, and technical skills (Haplin and Winer, 1957; Katz *et al.*, 1950; Katz and Kahn, 1951; Kahn, 1960; Mann, 1962).

In the present investigation the following dimensions of supervisory behaviour as suggested by Cammann *et al.* (1983) were studied :

Production orientation: The extent to which the supervisor stresses performance outcomes in dealing with sub-ordinates.

Control of work: The extent to which the supervisor maintains control of work that is being done by knowing the state of projects, planning work flows and making sure the work is being done correctly.

Work facilitation-Goal setting: The extent to which the supervisor helps the sub-ordinates to have clear and integrated goals so that they know what they should be doing.

Work-facilitation-problems solving: The extent to which the supervisor helps sub-ordinates solve work related problems.

Work-facilitation — sub-ordinate relations: The extent to which the supervisor maintains good communication and helpful, equitable relations with the sub-ordinates.

Bias: The extent to which the supervisor is biased by caste or sex in dealing with sub-ordinates.

Consideration: The extent to which the supervisor relates to and help sub-ordinates as individuals.

Participation: The extent to which the supervisor encourages sub-ordinates to participate in decisions.

Decision-decentralisation: The extent to which the supervisor makes important decisions by involving sub-ordinates.

Competence: Overall supervisory competence.

Job Technology

The term technology which comes from the Greek work 'techne' means art and it signifies knowledge, capacity or skill an individual uses to produce a product or artifact. As used in modern world technology refers to the application of knowledge for human purposes (Rousseau, 1983).

Organisational scientists have used the term technology to refer any thing from job routineness to the hardness of raw materials. Theoretically organisational technology has been defined in a number of ways. It has been defined in terms of technical complexity (Woodward, 1958), job complexity (Bell, 1967), mechanisms or processes by which an organisation turns out its products or service (Harvey, 1968) or routineness of work.

Hickson *et al.* (1969) defined technology in terms of techniques organisation uses in its work flow. Inkson *et al.* (1970) defined technology in terms of work flow integration, the degree of automated, continuous, fixed sequence of operations. Scott (1975) has argued that to study the technology of an organisation it is necessary to view it as a mechanism for transforming input into output.

is achieved when the ratio of rewards equals the ratio of merits (Birnbaum, 1983). The theoretical concept of equity has been applied to the problem of detecting salary inequalities by various researchers (Birnbaum, 1979, 1981; Roose and Doherty, 1978).

The following aspects of pay administration as suggested by Cammann *et al.* (1983) were investigated in the present study:

External equity: Degree to which the respondent's pay is equitable in comparison to the pay of comparable people in other related organisation.

Personal equity: The degree to which the respondent's pay is equitable given his/her own inputs on the job.

Performance-pay contingency: The degree to which the respondents perceive that pay is contingent on his/her level of performance.

Pay administration satisfaction: The degree to which the respondent is satisfied with how pay is administered in the organisation.

Pay satisfaction: The degree to which the respondent is satisfied with the amount of his/her pay.

Time Use

The effectiveness in use of time reflects the effectiveness in managing the activities. The concept of time use refers to the degree to which an individual in the organisation organises his time in order to complete the activities assigned to him. Time use and structuring of time of the employees in an organisation is an important factor which influences the productivity and efficiency of the employees.

Only very few studies have been reported on the time structuring of employees in the organisation. In a comparative study of use of time among the employed and unemployed people, Jahoda (1978) found that the sense of time of unemployed was disintegrated. She observed a blunted sense of time relations among the unemployed and a lack of purpose that affected their daily lives. It has been observed that characteristic of job play an important role in time use and structuring of time. Feathr and Bond (1983) while studying the time structure and purposeful activity among the employed and unemployed university graduates came out with the following four dimensions of time use :

(i) Engagement, (ii) Direction, (iii) Structure, and (iv) Routine. Jahoda (1979) also used the above four dimensions of time use in her study.

In the present study time use of the employees of the department of agriculture in the developed and the less developed state was studied with respect to the above four dimensions which are detailed below:

Engagement: It refers to the degree to which an individual has purpose, meaning and interest in his daily activities of work.

Direction: It refers to the degree to which an individual has direction, aim and less difficulties in initiating activities.

Structure: It refers to the degree to which an individual organises his work and does not remain in inertia due to lack of organisation.

Routine: It refers to the degree which an individual activities are planned and has routine.

Creativity Climate

The importance of creativity in the development of new ideas and products has been emphasised by Israeli (1946) and Karoles (1964). According to the above authors creativity is 'the process by which something new is produced'. The new idea or object must contribute to the solution of the problem. Guiford (1963) defined creativity as the 'combination of aptitude factors and disposition that enables a person to use his importance in novel ways'. O'Brien *et al.* (1963) defined creativity as the process of manipulating the environment, the environment which results in the production of new ideas, patterns or relationships.

The creativity of a person depends upon a number of factors such as personal traits, abilities and environment. The environment affect a person's attitude, traits and behaviour. The environment shapes though not determine the creativity. A punitive environment can result in fearfulness and blockage. On the other hand an environment which encourages experimentation may stimulate innovative effort (Khandwalla, 1984). For example, Thomas Alwa Edison was fortunate to have a mother who valued ingenuity in her child more than formal education (Untermeyer, 1955). "A home or academic or work culture characterised by commitment to experimentation and discontent with *status quo* may stimulated original thinking" (Khandwalla, 1984). A

Goal attainment model: The goal approach to organisational effectiveness defines organisational effectiveness in terms of extent to which the organisation achieves its goals or objectives (Price, 1968, 1972; Steers, 1975). Though the goal approach is most widely used by students of organisation, yet there is little agreement among the researchers on how to define goals and objectives. There are two major approaches for identifying the goals of the organisation, namely, 'prescribed goal approach' and 'derived goal approach'. The 'prescribed goal approach' (also known as goal approach) emphasises that the focus on formal charter of the organisation or on the top management as the most valid sources of information concerning organisational goals. Under the derived goal approach investigator derives the goal of the organisation from his functional theory thus arriving at goals independent of the intentions and awareness of the members. It is also known as functional approach (Yuchtman and Seashore, 1967).

The goal approach has been attacked on various grounds. For example, Katz and Kahn (1966) while noting that "the primary mission of an organisation as perceived by its leaders furnishes a highly informative set of clues", go on to point out the "nevertheless that stated purpose of an organisation is given by its by-laws or in the reports of its leaders can be misleading, such statements of objectives may idealise, ratinalise, distort, omit, or even conceal some essential aspects of the functioning of the organisation".

The functional approaches to organisational effectiveness can be characterised as 'normative' in the sense that the investigator reports what the goals of the organisation are or should be, as dictated by the logical consistency of his theory about the relationship among parts of larger social system (Yuchtman and Seashore, 1967). The functional or derived goal approach has advantage over the prescribed goal approach since it solves the problem of identifying the ultimate goal of complex organisation. However, the functional approach does not take organisation as frame of reference in identifying the goal. There is also disagreement on how to measure achievement assuming that goals have been measured in a satisfactory way. How are the progress towards accomplishing organisational goals be measured? Should they be indexed by employees ratings, supervisory ratings, or by analyst using presumably objective archival records? Defining objectives and measuring their achievement is critical, since effectiveness criteria can be designed to reflect the perspective of atleast three different reference groups profitability for the owners, satisfaction for the

employees, and societal value for the society at large (Coulter, 1979).

Processual model: Many organisational researches have defined the effectiveness construct in terms of one or more processual variables, which describe the organisation's internal operations and those that describe the link between the organisation and its environment (Coulter, 1979). The measures of internal processes include flexibility (Georgopoulos and Tannenbaum, 1957; Schein, 1970; Mott, 1972), open communications (Schein, 1970), man-power acquisition, retention and utilisation (Negandhi and Reimann, 1973), creativity (Schein, 1970) and adaptability (Bennis, 1971; Duncan, 1973; Webb, 1974). Researchers who have defined effectiveness in terms of the organisation's control over its environment (Katz and Kahn, 1966; Yuchtman and Seashore, 1967), and its ability to compete successfully in acquiring scarce and valued resources (Yuchtman and Seashore, 1967).

Behavioural-attitudinal model: According to this model it is stated that certain behavioural and attitudinal characteristics of individuals or groups of employees give the precise measure of an organisational effectiveness. The prominent characteristics which are used as effective ness measures as reported by Coulter (1979) are as follows : employees satisfaction (Friedlander and Pickle, 1968; Negandhi and Reimann, 1973), absence of tension and conflict within subgroups (Georgopoulos and Tannenbaum, 1957; Ghorpade, 1971), psychological commitment (Schein, 1970), turnover and absenteeism (Campbell, 1977), interpersonal relationship (Negandhi and Reimann, 1973), and morale (Price, 1968).

Approaches in Public Sector Organisations

Lot of work has been done on measuring the organisational effectiveness of public sector organisation (e.g. Mark, 1972; Hayes, 1978; Morris, Corbelt and Usilanr, 1972; Lucey, 1978). The following four approaches have been identified :

Input/output approach: Productivity is defined as efficiency (output compared to inputs) and effectiveness (output compared to some standard, usually cost), under this approach. The measures of performance are based on comparisons of output indexes, unit costs, work loads, activity rates, peakload requirements and safety standards. Calculations and applications of these indexes require definition and measurement the problems (Mott, 1972). The following dimensions of organisational adaptability in the department of agriculture were investigated in the present study :

Adaptability-anticipating problems: It refers to the abilities of members of organisational unit in anticipating future problems and preventing them from occurring it or minimising their effect.

Adaptability-awareness of potential solutions: It refers to the organisational abilities to keep abreast of new technologies and methods applicable to the activities of the organisation.

Adaptability-quickness in adopting new technologies: It refers to the degree to which the solutions to the problems in the form of new technologies are quickly adopted by the organisation.

Flexibility: Organisational flexibility which refers to the organisational ability to adjust its centres of power quickly to cope with temporarily unpredictable overloads of work that require significant but temporary modifications of roles by affected members (Mott, 1972).

Flexibility is conceptually different from adaptability because the organisational changes that result from meeting emergencies are temporary, usually the organisation returns to its pre-emergency structure. Adaptive changes are likely to be more permanent.

A Conceptual Model for the Study of Organisation

Figure 2.1 depicts the conceptual model of study the main classes of variables of organisations. It shows that organisational effectiveness is a product of work related behaviour of individuals and groups, with these in turn partially determined by prevailing attitudes, beliefs and expectations of individuals and of individuals clustered in work groups. The model also eludes the attributes of individual characteristic as well as organisational structure and job design, technology and other variables as determinants of overall individual behaviour and organisational effectiveness. The macro-organisational context is also portrayed as influencing the attitudes and behaviour of organisation members as well as the characteristic of the organisation.

Relevant Variables in the Assessment of Organisation

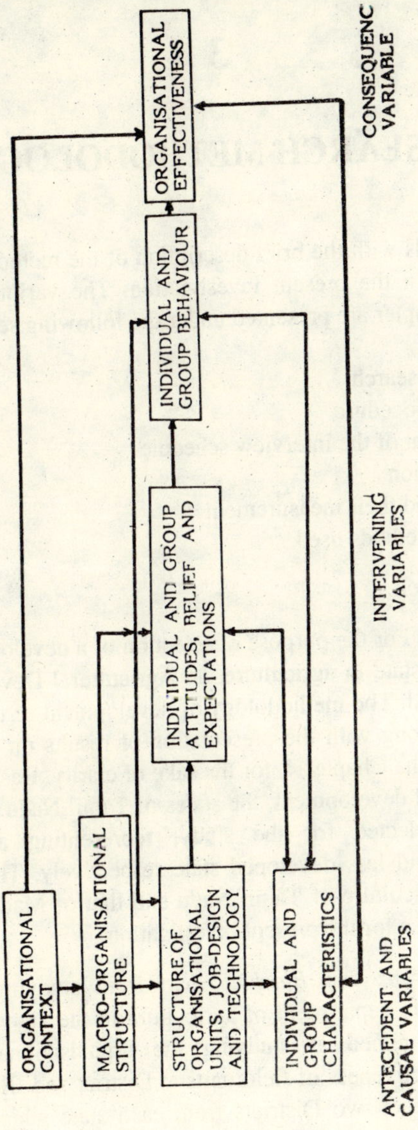

FIG. 2.1: Conceptual framework of the study

3

RESEARCH METHODOLOGY

This chapter deals with the brief description of the methods and procedures followed in the present investigation. The various aspects included in this chapter are presented under the following sections :

1. Locale of research
2. Sampling procedure
3. Development of the interview schedule
4. Data collection
5. Variables and their measurement
6. Statistical methods used.

Locale of Research

Selection of state: For the purpose of selection of a developed state and a less developed state in agriculture, an Agricultural Development Index was developed. The methodology of development of the Index has been discussed along with the presentation of results on the development of Index in the Chapter 4, for the sake of clarity. Based on the index of agricultural development, the states of Tamil Nadu and Madhya Pradesh were selected for the study, representing agriculturally developed state and less developed state respectively. Thus the state department of agriculture of Tamil Nadu and that of Madhya Pradesh became the universe for the present investigation.

Selection of Districts and Blocks: Apart from studying the headquarters of the state department of agriculture in the above two states the investigation aimed at studying the organisational structure, processes and effectiveness of field units at District and Block level. It was decided to select two Districts from each state which would be representatives of all the field units of the state department of agriculture. The Districts and the Blocks from each state selected for the study (Fig. 3.1) in consultation with the officials of the department of agriculture are given below:

Relevant Variables in the Assessment of Organisation

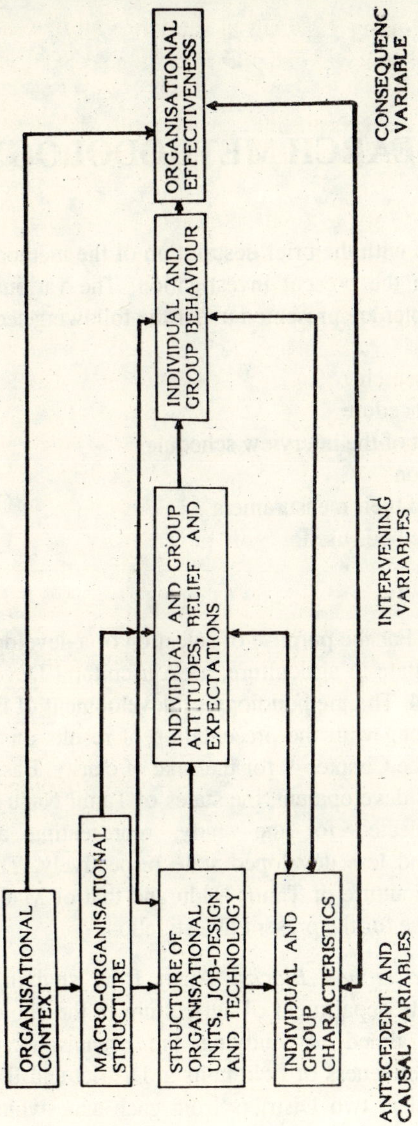

FIG. 2.1: Conceptual framework of the study

3

RESEARCH METHODOLOGY

This chapter deals with the brief description of the methods and procedures followed in the present investigation. The various aspects included in this chapter are presented under the following sections :

1. Locale of research
2. Sampling procedure
3. Development of the interview schedule
4. Data collection
5. Variables and their measurement
6. Statistical methods used.

Locale of Research

Selection of state: For the purpose of selection of a developed state and a less developed state in agriculture, an Agricultural Development Index was developed. The methodology of development of the Index has been discussed along with the presentation of results on the development of Index in the Chapter 4, for the sake of clarity. Based on the index of agricultural development, the states of Tamil Nadu and Madhya Pradesh were selected for the study, representing agriculturally developed state and less developed state respectively. Thus the state department of agriculture of Tamil Nadu and that of Madhya Pradesh became the universe for the present investigation.

Selection of Districts and Blocks: Apart from studying the headquarters of the state department of agriculture in the above two states the investigation aimed at studying the organisational structure, processes and effectiveness of field units at District and Block level. It was decided to select two Districts from each state which would be representatives of all the field units of the state department of agriculture. The Districts and the Blocks from each state selected for the study (Fig. 3.1) in consultation with the officials of the department of agriculture are given below:

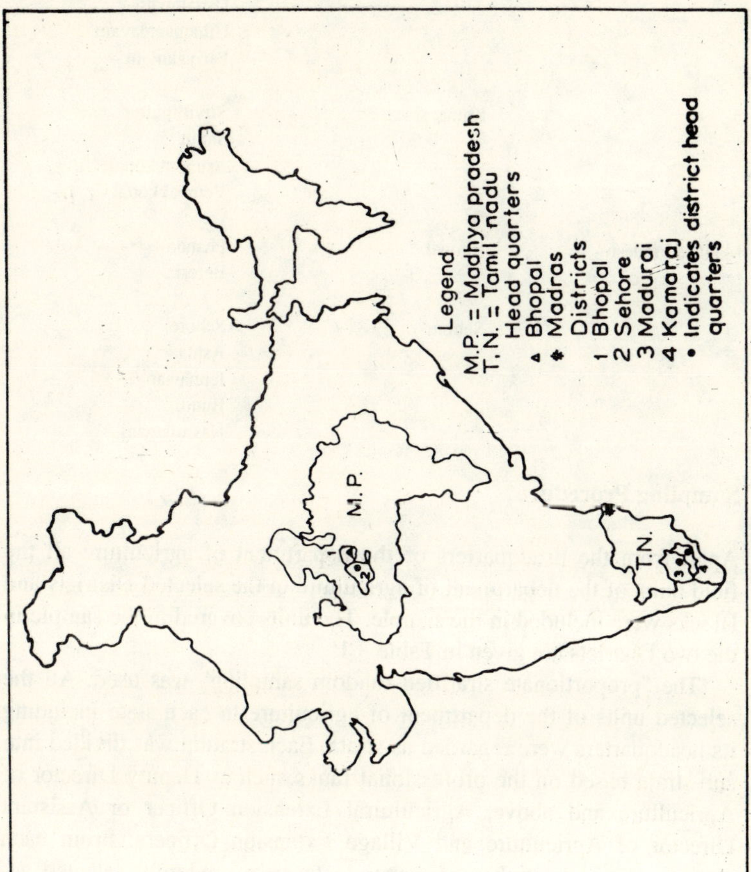

FIG. 3.1: Locations of the headquarters and district level offices of the department of agriculture

State	*Districts*	*Blocks*
Tamil Nadu	Madurai	Melur Thirumangalam Usilampatti Uthamapalayam Periyakulam
	Kamaraj	Srivilliputtur Sattur Aruppukkottai Vembakkottai
Madhya Pradesh	Bhopal	Phanda Beresia
	Sehore	Sehore Ashta Ichhawar Bundi Nasrullaganj

Sampling Procedure

Apart from the headquarters of the department of agriculture all the field units of the department of agriculture in the selected Districts and Blocks were included in the sample. The units covered in the sample in the two Districts are given in Table 3.1.

The "proportionate stratified random sampling" was used. All the selected units of the department of agriculture in each state including its headquarters were regarded as strata. Each stratum was divided into sub-strata based on the professional ranks such as Deputy Director of Agriculture and above, Agricultural Extension Officer or Assistant Director of Agriculture and Village Extension Officers. From each stratum required number of respondents were randomly selected according to the total number of employees of the unit. The heads of the field units and sections were selected purposively. The total number of respondents selected from the department of agriculture of Tamil Nadu and Madhya Pradesh according to their professional categories are given in Table 3.2.

Development of the Interview Schedule

The interview schedule was developed keeping in view the objectives of the study. Apart from information on demographic characteristics of the respondents, various information with regard to selected dimensions of organisational structure, processes and effectiveness were sought through the interview schedule. Most of scales or instruments that were required to measure the variables had been selected after an extensive search of literature and were used with slight modifications to suit the needs of the present investigation. The interview schedule used in the study has been given in Appendix I.

TABLE 3.1: Field units included in the sample

Districts		Field units
		TAMIL NADU
1. Madurai	1.	Office of the Joint Director of Agriculture, Madurai
	2.	Office of the Assistant Director of Agriculture and its block units at Melur, Thirumangalam, Usilampatti, Uthamapalayam and Periyakulam
	3.	Soil testing laboratory, Madurai
	4.	Pesticide testing laboratory, Madurai
2. Kamaraj	1.	Office of the Joint Director of Agriculture, Virudhunagar
	2.	Office of the Assistant Director of Agriculture and its block units at Sattur, Srivilliputtur, Aruppukkottai, Vembukkottai and Sivakasi
	3.	State seed farm, Devadanam
		MADHYA PRADESH
1. Bhopal	1.	Office of the Joint Director of Agriculture, Bhopal
	2.	Office of the Deputy Director of Agriculture, Bhopal and its field units at Phanda and Beresia
	3.	Office of the Agricultural Engineer, Bhopal
	4.	Office of the Assistant Soil Conservation Officer, Bhopal
	5.	Soil testing laboratory, Bhopal
	6.	Agricultural farm, Phanda
2. Sehore	1.	Office of the Deputy Director of Agriculture at Sehore
	2.	Office of Sub-Divisional Agricultural Officer at Sehore and Bundi and its field units at Sehore, Ashta, Ichhawar, Bundi and Nasrullaganj
	3.	Soil testing laboratory, Sehore
	4.	Office of Assistant Soil Conservation Officer, Sehore

TABLE 3.2: The distribution of sample respondents in the selected state departments of agriculture

State department of agriculture	Location	Frequency distribution of respondents			
		Deputy Director and above	Agricultural Extension Officer/Assistant Director or equivalent officer	Village Extension Worker	Total
Tamil Nadu	Headquarters	19	12	—	31
	District level	11	28	15	54
	Block level	—	10	55	65
Madhya Pradesh	Headquarters	20	15	—	35
	District level	10	22	18	50
	Block level	—	13	52	65
	Total respondents	60	100	140	300

Reliability of the scales: Through all the scales used in the study and developed by others had been tested for their validity and reliability, it was felt necessary to test the scales or instruments for their reliability. The reliability of all the scales was tested by the test-retest method. In this method, the instruments measuring various variables were administered twice on 15 days interval to 20 selected officials of the department of agriculture of Tamil Nadu belonging to different professional cadres but who did not come under the sample of the study. The two sets of scores obtained were correlated and the coefficients of reliability were found. The reliability coefficients of different scales or instruments are given in Table 3.3. All the coefficients were found to be statistically significant at 0.01 level of probability indicating that the scales or instruments used were highly stable with high internal consistency.

Validity: The validity of the scales and instruments especially developed for the study was attained through the content validity. The content validity is a validity through assumption (Guilford, 1965). The main criterion of the content validity is to make sure that the contents of the scale represent the subject matter which is important for the variable under study. The content validity was kept in mind and was ensured in the collection and selection of the items while developing the scales. Care was taken to include appropriate scale items to represent the variable that was to be measured.

The scoring technique: The scoring procedure as suggested by the authors of the scales or instruments was followed. Most of the scale items were either on five point continuum or seven point continuum with the scores ranging from 1 to 5 or 1 to 7. The negative statements were given the reverse scoring.

Data Collection

Data were collected from the respondents through personal interviews with the help of the structured interview schedules developed for the study. As the data were to be collected from different levels of officials of department of agriculture of Madhya Pradesh and Tamil Nadu, the interview schedule originally prepared in English was translated into Hindi and Tamil and got printed. The data were personally collected by the researcher himself during the period April 1986 to August 1986.

TABLE 3.3: **The reliability coefficients in test-retest method for the variables under study**

Sl. No.	Variables	No. of scale items	Reliability coefficients
1.	Units standardisation	5	0.81
2.	Distribution of unit authority	7	0.79
3.	Role interchangeability	3	0.92
4.	Unit incentives	2	0.78
5.	Unit sanction	2	0.74
6.	Job expertise	3	0.89
7.	Job standardisation	6	0.81
8.	Job authority	4	0.81
9.	Job pressure	5	0.92
10.	Job accountability	6	0.78
11.	Job feedback	6	0.77
12.	Expectation of rewards	3	0.82
13.	Expectation of sanctions	2	0.77
14.	Intra-unit communication flows	11	0.69
15.	Work group functioning	13	0.73
16.	Supervisory behaviour	25	0.84
17.	Job dependence among unit personnel	9	0.84
18.	Predictability of events	2	0.87
19.	Routineness of operation	2	0.91
20.	Job inputs adequacy	3	0.74
21.	Pay administration	12	0.91
22.	Time use	12	0.83
23.	Creativity climate	8	0.88
24.	Perception of organisational performance	10	0.91
25.	Job satisfaction	3	0.82
26.	Organisational involvement	4	0.78
27.	Adaptability	3	0.79
28.	Flexibility	1	0.85

Information from secondary sources like departmental documents, records, reports and other such sources were also collected for descriptive analysis. The information about the origin and history of the department of agriculture were collected from the historical documents on agriculture available in the libraries of Indian Agricultural Research Institute and Planning Commission.

Variables and Their Measurement

The procedure followed for the measurement of each variable is presented in Table 3.4. The conceptual definition of the each variable has been given in the chapter on Theoretical Orientation.

TABLE 3.4: Variables included in the study and their measurement

Variables studied	Operational definition	Method used
Organisational Context		
Origin and history	It refers to the origin and history of agricultural department of Tamil Nadu and Madhya Pradesh before and after Independence	Descriptive analysis, information collected from documents and reports
Budgeted expenditure	The plan expenditure of agricultural department for a specific period before independence and during Sixth Five Year Plan period	Descriptive analysis, information collected from documents and reports
Personnel Management Practices		
Recruitment	It refers to the procedures followed in recruitment of various cadres of employees of department of agriculture during 1985-86	Direct question
Salary	It refers to the total monthly salary of the employees of the department of agriculture during 1985-86	Direct question
Promotion	It refers to the promotional policies followed by the department of agriculture for different cadres of employees during 1985-86	Direct question
Macro Organisational Structure		
Size	It refers to the number of full time salaried employees of the department of agriculture during 1985-86	Information collected through direct question as well as from documents and reports
Number of supervisory level	It refers to the number of supervisors in the chain of command from Director of the department to the first line unit supervisor in the department of agriculture during 1985-86	Information collected through direct question as well as from documents and reports
Number of job titles	It is the number of job titles of technical staff as well as non-technical staff of the department of agriculture not counting the different grades within a service classification during 1985-86	Information collected through direct question as well as from documents and reports

TABLE 3.4 *(Contd.)*

Variables studied	Operational definition	Method used
Number of geographical units	It is the number of geographical operating units of the department of agriculture during 1985-86	Information collected through direct question as well as from documents and reports
Supervisor-staff	It is the ratio between the supervisory employees and the non-supervisory employees of the department of agriculture	It was computed by dividing the total number of supervisor by the total number of non-supervisory personnel
Percentage of supervisors above the bottom level unit supervisor	It refers to the percentage of all supervisors in the department of agriculture who are located hierarchically above the sub-divisional unit (above Assistant Director level)	It was computed by working out the percentage of supervisory positions above the Assistant Director level to the total number of supervisory position in the department
Technical employees-staff ratio	It is the ratio between the technical employees and the total non-technical employees of the department of agriculture	It was computed as the total number of technical employees divided by the total number of non-technical employees of the department
Ministerial* employees-staff ratio	It refers to the ratio between ministerial employees and the non-ministerial employee in the department of agriculture	It was computed as the total number of ministerial employees divided by the total number of non-ministerial employee in the department
Supporting employees-staff ratio	It refers to the ratio between supporting employees (class IV) and the total non-supporting employees of the department of agriculture	It was computed as the total number of supporting employees divided by the total number of non-supporting employees of the department

(Contd.)

* Ministerial employees refer to the administrative staff.

TABLE 3.4 *(Contd.)*

Variables studied	Operational definition	Method used
Percentage of posts filled into the total number of posts sanctioned	It refers to the percentage of technical posts filled into the total number of categories of technical posts sanctioned	It was computed by working out the percentage of technical posts filled into the total number of categories of technical posts sanctioned
Socio-Personal Variables		
Age	It refers to the chronological age of the respondents in completed years as on April 1986	Direct question
Family size	It refers to the respondent's total number of dependent members	Direct question
Total years of service	It refers to the respondent's total number of completed years of service in the department of agriculture	Direct question
Unit Structure		Index developed by Van de ven and Ferry (1980)
	The term unit refers to the organisational unit of the department of agriculture consisting of a supervisor and all personnel reporting directly to that supervisor at Divisional, District and Block level	
Unit standardisation	The degree to which work rules, standard operating procedures are specified in detail and written and also followed by unit personnel for coordination, control and evaluation of unit activities of the department of agriculture	
Supervisory authority	It is the perception of employees about the distribution of authority among unit supervisors of the department of agriculture on the unit decisions regarding task or work to be performed, rules and procedures to be followed and unit performance or appraisal criteria	

(Contd.)

TABLE 3.4 *(Contd.)*

Variables studied	Operational definition	Method used
Unit employee authority	It is the perception of employees about the distribution of authority among employees of an unit of the department of agriculture on the unit decisions regarding task or work to be performed, rules and procedures to be followed and unit performance or appraisal criteria	
External authority	It is the perception of employees about the distribution of authority among the higher management of the department of agriculture on the unit decisions regarding tasks or work to be performed, rules and procedures to be followed and unit performance or appraisal criteria	
Role interchangeability	It is the degree to which employees of the units of department of agriculture perform the same tasks and qualified in performing another's job. It also refers to the frequency of job rotation the employees have in the unit	
Unit incentives	Degree to which the groups' performance in the units of department of agriculture is rewarded in monetary and non-monetary terms	
Unit sanctions	Degree to which the members in the group will be reprimanded or told to 'shape up' to improve, because of failure to achieve the expected performance of the units of department of agriculture	
Job Design		Index developed by Van de ven and Ferry (1980)
Job standardisation	Degree to which the roles and task assignments that make up a job are written out in a job description in the department of agriculture. It refers to the number and detail of rules and procedures established to guide the job incumbent in work performance	

(Contd.)

TABLE 3.4 *(Contd.)*

Variables studied	Operational definition	Method used
Job expertise	It refers to the skill level of employees of the department of agriculture in terms of their formal education, length of job entry orientation and the amount of time spent by the employees in on the job training and reading necessary for upgrading and remaining current in the knowledge needed to perform the job	
Job authority	Amount of discretion or influence that an employee of the department of agriculture exercises in making job related decisions regarding (1) tasks and role to be performed, (2) procedures to be followed in doing the work and (3) handling work exceptions and problems	
Job pressure	It refers to the employees' amount of work load, time available to perform it and the extent to which the employee of the department of agriculture can control the pace of the work	
Job accountability	Degree to which an employee of the department of agriculture feels responsible and is asked to answer for his or her work decisions, achieving and appraising work standards and take credit or blame for the work	
Job feedback	Degree to which an employee of the department of agriculture receives information about the procedures and results of his/her job from the job itself, from other employees or colleagues, and from the supervisor	
Expectation of rewards	Degree to which an employee of the department of agriculture anticipates that good job performance will result in some reward like formal promotion or informal recognition	
Expectation of sanctions	Degree to which the employees of the department of agriculture anticipates that poor job performance will result in some punishment either informal reprimand or formal demotion	

(Contd.)

TABLE 3.4 *(Contd.)*

Variables studied	Operational definition	Method used
Intra-Unit Communication Flows		Index developed by Van de ven and Ferry (1980)
Written communication flows	Written information/communication flows are work related messages sent among unit employees of the department of agriculture through written memos, reports and letters	
Personal communications flows	Personal communications/information flows are work related messages sent among unit employees of the department of agriculture through normal one-to-one discussions	
Group communication flows	Group communications/information flows are work related messages sent among unit employees of the department of agriculture through group or staff meetings among three or more unit employees	
Work Group Functioning		Scale developed by Cammann *et al.* (1983) and suitably adopted for the present investigation
Group homogeneity	Degree to which members of the same work group in the department of agriculture are similar in their educational and technical training and socio-economic backgrounds	
Group goal clarity	Degree to which the members of the work group in the department of agriculture are clear about the job or work to be done and the aim and purpose of the group	
Group cohesiveness	Degree to which the members of the work group in the department of the agriculture feel that they are part of team and look forward to be with the team each day	

(Contd.)

TABLE 3.4 *(Contd.)*

Variables studied	Operational definition	Method used
Open group process	Degree to which the work group members of the department of agriculture express their feeling and real views about their job or work related matters. It also involves the extent of involvement or participation of each member in the decision of group with regard to work	
Internal fragmentation	Degree to which the members of the work group of the department of agriculture do not have respect for each other and involves in constant bickering which separate the members from each other	
Supervisory Behaviour		Scale developed by Cammann et al. (1983) and suitably adopted for the present investigation
Production orientation	Degree to which the supervisor of the department of agriculture stress its employees to work hard and produce quality work with regard to the job assigned	
Control of work	The extent to which the supervisor of the department of agriculture controls the work through advance planning of work, maintaining high standard of work performance and handling the administrative work efficiently	
Work facilitation goal setting	The extent to which the supervisor of the department of agriculture helps the employees of his unit in having a clear cut defined goal related to the work. It also involves making the employees knowing what works have to be done and how should they be done	

(Contd.)

TABLE 3.4 *(Contd.)*

Variables studied	Operational definition	Method used
Work facilitation problem solving	The extent to which the supervisor of the department of agriculture helps the employees of his unit in solving work related problems and discover problems before they get too bad	
Work facilitation subordinate relations	The extent to which the supervisor of the department of agriculture kceps the employees of his unit informed of the latest development of the unit; helps the employees in developing their technical skills; deals with the employees in a fair manner; avoids showing favouritism based on caste or religion; and freely exchanges the ideas and feelings about the job related matters	
Bias	The extent to which the supervisor of the department of agriculture is biased towards the employees of the unit based on caste, religion, sex or region	
Consideration	The extent to which the supervisor of the department of agriculture helps the employees of his unit and is concerned about the employees general well-being and feels that every employee is important as an individual	
Participation	The extent to which the supervisor of the department of agriculture encourages the employees of his unit to participate in making work related decisions and encourages the employees to express their views or to speak when they disagree with a decision	
Decision-decentralisation	The extent to which the supervisor of the department of agriculture makes work related decisions by involving and consulting the opinions of the employees of the unit	

(Contd.)

TABLE 3.4 *(Contd.)*

Variables studied	Operational definition	Method used
Competence	It is the extent to which the supervisor of the department of agriculture is perceived by the employees of his unit as competent in his work and over all administration	
Job Technology		Scale developed by Van de ven and Ferry (1980)
Job dependence among unit personnel	Degree to which an employee's cycle of work (input, transformation and output) activities depend upon the activities performed by the supervisor, other members and people in other units or departments	
Predictability of events	Degree to which the employees of the department can anticipate and predict the nature of events which causes their work	Scale developed by Lynch (1974) and suitably adopted for the present investigation
Routineness of operations	Degree to which the work of the employees of the department of agriculture is characterised by routine and repetitions of procedures or methods to be followed in the work from one day to another	
Job inputs adequacy	Degree to which the inputs like information, materials and other resources are readily available to begin the normal work of the employees of the department of agriculture	
Pay Administration		Scale developed by Cammann *et al.* (1983) and suitably adopted for the present investigation

(Contd.)

TABLE 3.4 *(Contd.)*

Variables studied	Operational definition	Method used
Pay administration satisfaction	The degree to which the employees of department of agriculture are satisfied with the general administration of pay (payment in time) and with the policies regarding annual increment, dearness allowance, travelling allowance etc.	
Pay satisfaction	The degree to which the employees of the department of agriculture feel happy about the total amount of pay they make	
Performance-pay contingency	The degree to which the employees of the department of agriculture perceive that their pay depends upon how well they perform their job	
Personal equity	The degree to which the pay of the employees of department of agriculture is perceived to be fair considering the skill, type and amount of work done by them	
External equity	Degree to which the pay of employees of the department of agriculture is equitable in terms of total emoluments in comparison to the comparable cadres in other developmental departments like health, education, public works etc.	
Time Use		Scale developed by Feather and Bond (1983) and suitably adopted for the present investigation
Direction	Degree to which an employee of the department of agriculture has clear cut ideas about what he will do next, and finds his work not being hampered by unscheduled activities	

(Contd.)

TABLE 3.4 *(Contd.)*

Variables studied	Operational definition	Method used
Engagement	It is the degree to which an employee of the department of agriculture has interest, liking, purpose in his work as well as derives satisfaction out of his work	
Structure	Degree to which an employee of the department of agriculture organises his work and does not find his time slipping away or remain inactive due to lack of organisation	
Routine	It is the degree to which an employee of the department of agriculture plans his day-to-day work or has daily routine of work and has a rough idea or plan about the future work	
Creativity Climate		Scale developed by Khandwalla (1984) and suitably adopted for the present investigation
Degree of stimulation	Degree to which an employee of the department of agriculture has opportunities to learn new techniques or methods with regard to his work	
Availability of rewards	It refers to the availability of monetary or non-monetary awards to the employees of the department of agriculture for any innovative ideas or new ways of doing things related to performance of their work	
Availability of constructive feedback	It refers to the degree to which constructive suggestions and evaluations are available for the innovaative ideas or doing things related to the work for an employee of the department of agriculture	

(Contd.)

TABLE 3.4 *(Contd.)*

Variables studied	Operational definition	Method used
Freedom of thought	It refers to the degree to which an employee of the department of agriculture has freedom to hold positions or beliefs with regard to the activities of the department, which is different from that of his superior or headquarters	
Absence of premature criticism	Degree to which an employee of the department of agriculture is not criticised prematurely or punished for his/her novel idea or new ways of doing things with regard to the work	
Opportunities for higher learning	Degree to which an employee of the department of agriculture has opportunities for higher level of training on a regular basis as per the need of the job	
Availability of diverse view points	Degree to which an employee of the department of agriculture is exposed to different intellectual views or ideas or interaction concerning his work. It also refers to the degree to which an employee need not conform to the office norms and procedures in needed circumstances	
Organisational Effectiveness		Rating scale developed, as suggested by Georgopoulos and Mann (1962)
Perception of organisational performance	It is the degree to which the employees of the department of agriculture perceive its performance with regard to its major goals or activities. The following are the activities against which the employees were asked to rate the performance : (1) Reputation for work excellence among the farmers, (2) quality of work done by the unit/ department, (3) total quantity of work done, (4) achievement of target, (5) provision of advisory service to	

(Contd.)

TABLE 3.4 *(Contd.)*

Variables studied	Operational definition	Method used
	farmers, (6) supply of inputs or procurement of seed materials, (7) training of farmers, (8) training of the employees, (9) conducting adaptive trials and Minikit programme, (10) regularity in field visit. The above dimensions of performance of department of agriculture were included in the study after consulting with the officials at various levels of the department	
Job satisfaction	It refers to how an employee of the department of agriculture is happy or satisfied with work environments, co-workers and overall aspects of the job	Scale developed by Cammann *et al.* (1983)
Organisational involvement	It refers to the extent to which an employee of the department of agriculture is personally concerned about activities and programmes of the department and considers it important for him	Scale developed by Cammann *et al.* (1983)
Adaptability — Flexibility		Scale developed by Mott (1972)
Adaptability — Anticipating problems	It is the degree to which the members of the department of agriculture anticipate future problems concerning agricultural production or development and takes measures to solve them	
Adaptability — Awareness of potential technology	It is the extent to which department of agriculture is aware about the new farm technologies, which are relevant to the overall increase in the agricultural production or productivities of the state	
Adaptability — Quickness in adopting new technology	It is the degree to which the department of agriculture is relatively quick in popularising and adopting new agricultural technologies in their production programmes	
Flexibility — Coping with emergencies	The degree to which the department of agriculture cope up with unforeseen situations like drought, floods, or other emergencies	

Statistical Methods Used

The statistical analysis such as '*t*' test, Pearson's coefficient of correlation (*r*) and Discriminant function were used in the study.

4

DEVELOPMENT OF A COMPOSITE AGRICULTURAL DEVELOPMENT INDEX

The difference in the levels of agricultural development of various states of India has been a great concern to the planners, policy-makers and to the scientists. Despite the planned development of more than three decades the inter-state or regional disparity has not been reduced. Though the over all progress of agricultural development in India especially after mid-1960s has been quite impressive and resulted in green revolution, the fruits of it have not been shared evenly by various states or regions of India.

Continuous efforts have been made by scientists, planners and other interested personnel to measure the levels of development of various regions in order to classify the regions and know where a given regions stands in relation to others. There is a need to know the position or level of agricultural development of various regions of India since our economy depends mainly on agriculture. Inter and intra-regional comparison of the level of agricultural development will help in the planning for reducing imbalances. The less developed regions can greatly learn from the experiences of developed regions and can make concerted efforts on the needed areas of agricultural development. The classification of regions on the basis of agricultural development is of special interest to the scientists who want to study the factors responsible for the differences in the development.

It is on the above context an attempt has been made to develop a composite index of agricultural development for the twenty-one states of India which may form a basis of comparison between the states. The primary objective of development of the index is to select and compare an agriculturally developed state with an agriculturally less developed state with regard to various dimensions of agricultural development administration. In other words the present study aims to study the differences (if any) between the structure, processes and effectiveness of agricultural development administration in a developed and a less developed state. Such comparison will reveal whether the developed and the less developed state differ with regard to various dimensions of agricultural development administration and will pin-point the areas in which improvements are needed.

The published literature on the earlier attempts to measure the level of agricultural development of various regions or states are not available though there are considerable amount of literature on the measurement of overall economic development. Several attempts have been made to develop the index of economic development. Bennet (1951) constructed a non-monetary index of development to focus the attention of international disparities. Census Commission of India has classified all the Districts of India into four categories (I to IV) in the ascending order of their development based on subjective weighting. The methodology of above classification as reported by Dasgupta (1971) was to arrange the Districts in quartiles in order of their observed values for 35 indicators of development and then to divide each quartile into two equal halves. The eighth values for each indicator are then added up for each District and finally the Districts are arranged into four categories in order of their development, or the basis of their total rank scores. Category 'I' is the lowest and category 'IV' is the highest, in order of development.

Dasgupta (1971) classified 294 Districts of India according to their degree of development based on 24 indicators of developments as given in the 1961 Census data. The statistical methods used in his analysis were as follows:

(i) First the correlation matrix of the socio-economic variables was examined to weed out the less significant ones and to find a smaller set of variables which moved together and which approximately stood for economic and social development;

(ii) Secondly, a principal component analysis was carried out on the basis of both this smaller set as also the original set of variables and the districts were classified into four development categories according to the values of the principal components;

(iii) Thirdly the discriminant analysis was applied to subject the classification into a more vigorous test. The results from these two analyses were than compared with each other; and

(iv) Lastly, on the basis of the comparison the results were summarised with new rank values for the districts.

Rao (1973) used six indicators in constructing an index of development of various regions in India. He developed a composite index of development based on covariation in various indicators of development. Other works in the above area have been done by Mukherjee (1980) and Nanjappa and Sudharshan (1981).

Attempts have also made by Planning Commission to study regional variations. In mid-Sixties Planning Commission took up a study on regional variation in social development and levels of living.[1] Emphasis was given to reduce regional inequalities in Fourth Plan through special programmes for dry farming and desert areas and small farmers. During Fifth Plan period Chakravarty Committee on Backward Area Development was appointed by Planning Commission. This committee did an index based exercise for the identification of backward Districts, based on the following indicators[2] :

1. Density of population per square kilometre.
2. Percentage of agricultural workers to total working force.
3. Gross value of output of foodgrains per head of rural population.
4. Gross value of output of all crops per head of rural population.
5. Gross value of output of non-food grains per head of rural population.
6. Percentage of total establishment using electricity to total number of establishments (manufacturing and repair).
7. Percentage of household establishments using electricity to total household establishments.
8. Percentage of non-household establishments using electricity to total non-household establishments.
9. Number of workers in registered factories per lakh of population.
10. Length of surfaced roads per lakh of population.
11. Percentage of male literates to male population.
12. Percentage of female literates.
13. Percentage of total literates.

Using the index approach with the above indicators of the development Chakravarty Committee identified 179 backward districts of India.

Alagh and Bhalla (1979)[3] have classified various districts of India into high, medium and low productivity districts based on the value of agricultural output per hectare. The high, Medium and low productivity district had a value of output per hectare (in Rupees), as above

1300, between 700 to 1300 and less than 700 respectively. Out of 282 districts studied, 69 fell into the category of high productivity, 130 into medium category and 83 into low productivity. Further Alagh and Bhalla (1979)[4] also classified districts into high, medium, low and decelerating growth districts based on annual growth rate. The growth rate (in percentage) of each of the above categories of district was 4.5, 1.5 to 4.5, 0 to 1.5 and negative growth respectively.

Taxonomic Method of Measuring the Level of Agricultural Development of Various States of India

In the present study, the level of agricultural development of various states of India has been measured by Taxonomic method as reported by Srivastava (1983). The method of Taxonomy was developed by a group of Polish mathematicians in early 1950. During 1968, this method was recommended to UNESCO as a tool for classifying and comparing countries by their levels of development. And this method has been recently used to measure the levels of development and components of development of developing countries in Asia, Latin America, Middle-East and Africa. Various steps involved in the computation of composite index of agricultural development of the various states are given below :

Selection of indicators of development: The indicators of development for all the regions have to be selected based on theoretical considerations and inter-relations among the indicators. These indicators are to be arranged on a matrix form. It is necessary to classify the selected indicators into positive and negative ones, those which promote and those which retard development.

Let a set of '*n*' points represent region 1, 2, ... *n* for a group of indicators 1, 2, 3, ... m. These can be represented by a matrix as follows :

$$\begin{bmatrix} X_{11} & X_{12} \cdots\cdots & X_{1m} \\ X_{21} & & X_{2m} \\ \cdot & & \\ \cdot & & \\ \cdot & & \\ X_{n1} & & X_{nm} \end{bmatrix}$$

Taxonomic method aims at constructing a unitary index of development and it is necessary to add up the values of selected indicators at appropriate stage. This calls for converting the original values of the indicators into some standardised values by the following formula :

$$Z_{ij} = \frac{X_{ij} - \bar{X}_j}{S_j^2} \qquad \begin{array}{l} i = 1, 2, \ldots\ldots n \\ j = 1, 2, \ldots\ldots m \end{array}$$

where, $X_j = \Sigma X_{ij}/n$, and
$S_j^2 = (X_{ij} - X_j)^2 / n$

X_{ij} is the value of jth indicator for the oth region. These standardised values are further arranged in another matrix as follows :

$$\begin{bmatrix} Z_{11} & Z_{12} & \ldots\ldots & Z_{1m} \\ Z_{21} & Z_{22} & \ldots\ldots & Z_{2m} \\ \cdot & & & \\ \cdot & & & \\ \cdot & & & \\ \cdot & & & \\ \cdot & & & \\ \cdot & & & \\ Z_{n1} & & & Z_{nm} \end{bmatrix}$$

The next step involved is to find out the 'distance' from each region to another region for each of the standardised value. Thus it is necessary to convert several distances of each region into one single mathematical expression so that it would be easy to compare all the regions. When the values, thus obtained arranged in a matrix form, we get systematic distance matrix.

The elements of D_{ij} of the distance matrix is defined as:

$$D_{ij} = + \sqrt{\sum_{k=1}^{m} (Z_{ik} - Z_{jk})^2}$$

where $D_i = 0$ and $D_{ij} = D_{ji}$

D_{ij} is the square root of sum of squares of the elements j^{th} row to the corresponding elements of the jth row. Matrix is given by :

$$\begin{bmatrix} 0\ D_{12}\ D_{13} \ldots\ldots\ldots\ldots\ldots\ldots D_{1n} \\ D_{21}\ 0\ D_{23} \ldots\ldots\ldots\ldots\ldots\ldots D_{2n} \\ D_{31}\ D_{32}\ O \ldots\ldots\ldots\ldots\ldots\ldots D_{3n} \\ \cdot \\ \cdot \\ \cdot \\ \cdot \\ \cdot \\ \cdot \\ \cdot \\ D_{n1}\ D_{n2}\ D_{n3} \ldots\ldots\ldots\ldots\ldots\ldots 0 \end{bmatrix}$$

In each row there will be one point with the shortest or minimum distance (C_i) at the corresponding point with row that is :

$$C_i = \min (D_{ij}) \text{ and } C_i \neq 0$$

In the words of Srivastava (1983) "the first row in the distance matrix represents the composite distances between first region and every other regions and so on. In each row there will be one region with the shortest or the minimum distance to the corresponding region in the row. The former region is called the primary model of the latter and the latter the shadow of the former. One can hypothesise a situation when in one row there may be two or more regions with same shortest or same minimum distance to the corresponding regions in that row. However, the probability of such situation arising is virtually zero. Therefore, one can assume that there will always be only one region with shortest distance in each row".

The next step is to find out the pattern of measure of development of each region. An ideal region (C) is created, first by taking the highest or best standardised values of indicators of development of the groups of regions under consideration. The ideal region will have the highest values of positive indicators and the lowest values of negative indicators of development. A model region is simulated from a group of regions as it is unlikely that only one region in the group will have the highest or lowest value of the indicators. The distance between the ideal region and the every region in the matrix for each standardised value of selected indicators are calculated. These distances are arranged in another matrix, from which one can find out the pattern and measure of the development of each region. The pattern of development is simply the composite distance from the ideal region to other region and the measure of development is a function of pattern of development and critical distance from the simulated ideal region.

The pattern of development is obtained by following formula :

$$C_{io} = \sum_{k=1}^{m} (Z_{ik} - Z_{ok})^2$$

where C_{io} denotes the pattern of development (i = 1, 2, ... n) and Z_{ok} is the highest or the best standardised value.

The measure of development is given by :

$$d_i = C_{io} / C_o$$

where, $C_o = C_{ij} + 2 S_{io}$

$$\bar{C} = \sum_{1}^{n} C_{io} / n$$

$$S_{io} = \sqrt{\Sigma (C_{io} - \bar{C})^2 / n}$$

The measure of development is always non-negative and it lies between 0 and 1 though it may sometimes exceed 1 with a chance to near zero. The measure of development which is near to zero indicates that the region is developed and the measure of development which is near 1 indicates that the region is less developed.

Selection of Indicators of Agricultural Development

The following indicators which reflects the overall agricultural development were selected for the construction of the index of agricultural development based on the theoretical consideration and upon the opinion of experts in this field ;

1. Gross value of agricultural output (in Rupees) per hectare of net area sown (average for the years 1979-80 to 1981-82).
2. Per capita rural income in rupees (average for the years 1979-80 to 1981-82).

3. Per hectare yield of major cereal crops (rice, *jowar, bajra,* wheat, *gram* and *tur*) (average for the years 1978-79 to 1982-83).
4. Consumption (kg/ha) of nitrogen, phosphorus and potash per hectare of gross cropped area (average for the years 1978-79 to 1982-83).
5. Per hectare consumption (grams/ha) of pesticides (average of the years 1978-79 and 1982-83).
6. Percentage of net irrigated area to net cultivated area (average for the years 1977-78 and 1980-81).
7. Percentage of area under high yielding varieties for rice, wheat, *jowar, bajra* and maize (average for the years 1979-80 to 1982-83).
8. Percentage of village accessible to all weather roads (value for the year 1981).

Correlation Matrix of the Variables

Table 4.1 shows the inter-correlation matrix of the eight variables selected for the study. The results of correlation shows that there is a high degree of inter-correlation among the variables.

The value of crop output was positively and significantly related with yield of crops and all weather roads. It did not show any significant relationship with other indicators. Thus it is clear that the increase in yield of crops and marketing facilities (all weather roads) will result in increased value of crop output. The per capita rural income was positively and significantly associated with yield of crops, area under high-yielding varieties, fertilizer consumption and percentage of net irrigated area. This means that an increase in the area under high-yielding varieties, fertilizer consumption and the percentage of net irrigated area will result in the increase of per capita rural income. The yield of major crops was positively and significantly correlated with all indicators like area under high yielding varieties, fertilizer consumption, pesticide consumption, irrigated area and all weather roads.

Area under high-yielding varieties was positively and significantly associated with fertilizer consumption, pesticide consumption and net irrigated area. It is clear that an increase in irrigated area results in the increase in area under high varieties. The increase in the area under high-yielding varieties results in the increase of fertilizer as well as pesticide consumption. The fertilizer consumption was positively and significantly correlated with pesticide consumption, net area irrigated, and all weather roads. This means that an increase in the irrigated area

TABLE 4.1: Inter-correlation matrix of indicators of agricultural development

Indicators	1	2	3	4	5	6	7	8
1. Value of crop output	1.0000							
2. Rural income	0.0912	1.0000						
3. Yield of crops	0.6437**	0.6346**	1.0000					
4. Area under HYVs	0.3330	0.6944**	0.7818**	1.0000				
5. Fertilizer consumption	0.2008	0.6945**	0.7692**	0.8299**	1.0000			
6. Pesticide consumption	0.1554	0.2528	0.3875*	0.5287**	0.6630**	1.0000		
7. Net irrigated area	0.2353	0.5583**	0.7054*	0.6044*	0.7099**	0.3374	1.0000	
8. All weather roads	0.5005**	0.3501	0.6416**	0.3769	0.4843*	0.3533	0.3402	1.0000

* Significant at 0.05 level of probability

** Significant at 0.01 level of probability

as well as marketing facilities results in the increased consumption of fertilizers. The increase in the use of fertilizer also results in the increased use of pesticides.

Distance matrix

The inter-state agricultural development distance has been given in Appendix II. The first row in the distance matrix represents the composite distances between the first region and the every other region and so on. From the shortest distance in each row we can find out the primary model of agricultural development pertaining to that state. The primary model of each state is given below :

State	*Primary model*
1. Andhra Pradesh	Gujarat
2. Assam	Meghalaya
3. Bihar	Karnataka
4. Gujarat	Karnataka
5. Haryana	Jammu and Kashmir
6. Himachal Pradesh	Bihar
7. Jammu and Kashmir	West Bengal
8. Karnataka	Maharashtra
9. Kerala	West Bengal
10. Madhya Pradesh	Orissa
11. Maharashtra	Karnataka
12. Manipur	Meghalaya
13. Meghalaya	Assam
14. Nagaland	Bihar
15. Orissa	Madhya Pradesh
16. Punjab	Haryana
17. Rajasthan	Madhya Pradesh
18. Tamil Nadu	Jammu and Kashmir
19. Tripura	Himachal Pradesh
20. Uttar Pradesh	Bihar
21. West Bengal	Jammu and Kashmir

The primary model of each state is that state with which the state concerned maintains a minimum composite distance in agricultural development.

Measure of Agricultural Development

On the basis of measure of agricultural development all the 21 states have been ranked from developed to less developed in Table 4.2. (Fig. 4.1)

It can be seen from Table 4.2 and Fig. 4.1 that agriculturally most developed state in India was found to be Punjab with a measure of 0.1897, which is quite close to zero which is indicative of the fact that the overall agricultural development of Punjab is quite high when compared to other states. In contrast the least agriculturally developed state was Madhya Pradesh with a measure of 0.8683 which is quite-quite close to 1, indicating that the overall agricultural development of Madhya Pradesh is very poor when compared to other states of India. This shows a greater degree of disparity in agricultural development between various states of India and calls for greater attention to be paid to the less developed state.

Next to Punjab, the states in the order of agricultural development were Haryana (0.4656), Tamil Nadu (0.5088), Andhra Pradesh (0.5343), Jammu and Kashmir (0.5533). These states had a measure of development ranged from 0.4656 to 0.5533, and these measures were

TABLE 4.2: Measures of agricultural development of states

Rank	State	Measure of Agricultural Development*
1.	Punjab	0.1897
2.	Haryana	0.4656
3.	Tamil Nadu	0.5088
4.	Andhra Pradesh	0.5343
5.	Jammu and Kashmir	0.5533
6.	Kerala	0.6202
7.	West Bengal	0.6467
8.	Gujarat	0.6495
9.	Uttar Pradesh	0.6622
10.	Manipur	0.6966
11.	Karnataka	0.7183
12.	Himachal Pradesh	0.7443
13.	Bihar	0.7530
14.	Maharashtra	0.7671
15.	Meghalaya	0.7677
16.	Tripura	0.7746
17.	Assam	0.7794
18.	Nagaland	0.8289
19.	Orissa	0.8338
20.	Rajasthan	0.8658
21.	Madhya Pradesh	0.8683

*Closer the measure of development to '0' the more developed is the region

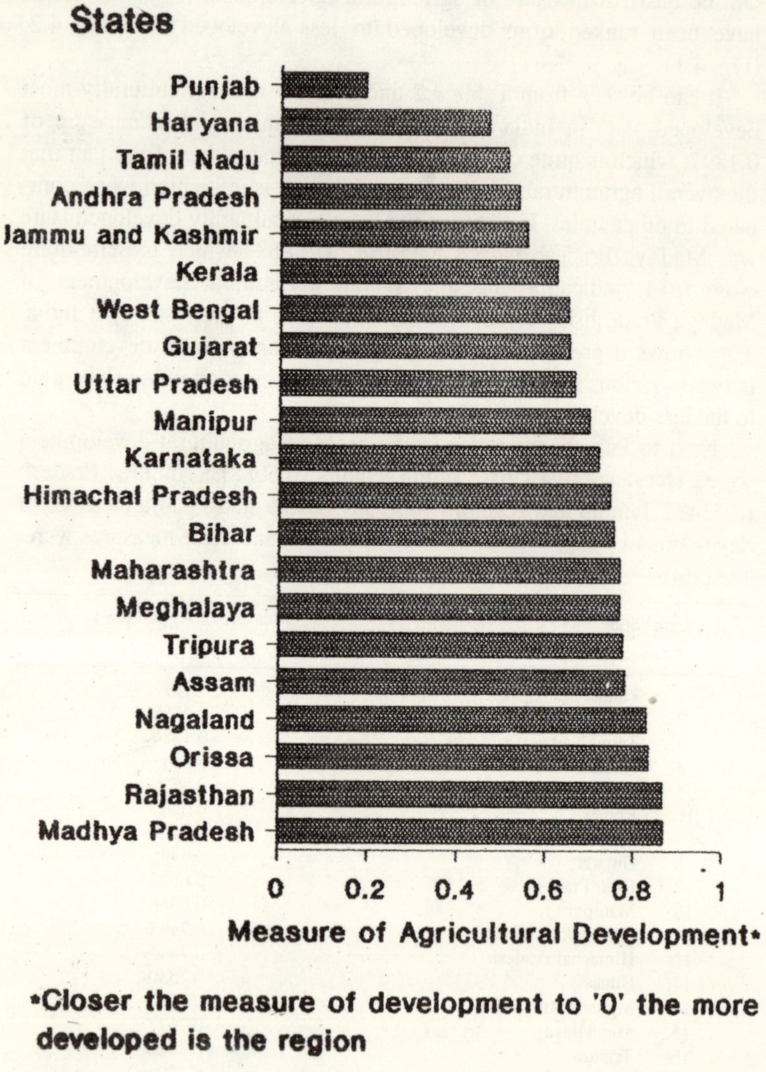

FIG. 4.1: Measure of agricultural development of states

quite close to each other. These states can be said to be relatively developed in agriculture. The Haryana which came next to Punjab had a development measure of 0.4656 which is quite far from the measure of Punjab i.e. 0.1897. This indicates that Punjab stands well ahead of all other states not only in comparison with less developed state but also with relatively developed states with regard to agricultural development. A long-term and a massive investment efforts are required by all other states especially the states at the bottom of agricultural development to catch up with the development of Punjab.

The states like Rajasthan (0.8658), Orissa (0.8338), Nagaland (0.8289) had very high measures (above 0.8) and can said to be in the bottom of agricultural development in India. And these states need special attention to improve their status in agricultural development.

The states like Assam (0.7794), Tripura (0.7746), Meghalaya (0.7677), Maharashtra (0.7671), Bihar (0.7530), Himachal Pradesh (0.7443), Karnataka (0.7183) and Manipur (0.6966) also had a high measure of development (above 0.69 and up to 0.79). These states can be said to be in the relatively lower level of agricultural development. It is a point to note here that all the north-eastern states have lower level of agricultural development.

The states like Kerala (0.6202), West Bengal (0.6467), Gujarat (0.6495) and Uttar Pradesh (0.6622) had a relatively lower score (0.6 to 0.68) and can be placed in the medium position in agricultural development.

From the examination of the results it could be stated that the findings supported the general hypothesis (GH-1), namely it is possible to construct an agricultural development index of the states and arrange them in the order of agricultural development.

REFERENCES

1. Regional Variations in Social Development and Levels of Living — A Study of the Impact of Plan Programmes (1967), Planning Commission, New Delhi.
2. As reported in Report on General Issues Relating to Backward Area Development (1981), National Committee on the Development of Backward Areas, Planning Commission, New Delhi.
3. Alagh, K.A. and Bhalla, G.S. (1979). *Performance of Indian Agriculture*, Sterling Publishers, New Delhi.
4. *Ibid.*

5

ORGANISATIONAL CONTEXT

Based on the agricultural development index developed for various states, the departments of agriculture of Tamil Nadu and Madhya Pradesh representing the developed state and the less developed state respectively were selected purposively for the study. Two organisational contextual factors namely origin and history and the budgeted expenditure of the above departments of agriculture were studied in the present investigation. The results of the study are presented below:

Origin and History of the Agricultural Department of Tamil Nadu*

The present state of Tamil Nadu was a part of the erstwhile Madras Presidency which also consisted of a major part of the present state of Andhra Pradesh and Kerala. The total geographical area of the Province was 1,42,277 square miles with a population of 41.40 million during 1911. The population rose to 46.74 million during 1931. During 1931 Madras Province had 26 Districts with 340 towns and about 51,500 villages. The net area sown during 1933-34 was 33.88 million acres and about 29 per cent of the area was irrigated. The first recorded attempts in agricultural improvements is the introduction of a variety of cotton from Mauritius in 1790, the relic of which is the Bourbon cotton. In 1835, an Agri-Horticultural society was founded in Madras to promote agricultural improvements. In 1854, an agricultural exhibition and a cotton show were arranged and prizes were awarded.

Practical interest in agricultural development was awakened as early as 1863 by the then Governor who drew attention to the prevalent practice of continuous cropping, deficiency of nutrients in soil, defective implements and lack of accurate statistics in the state. As a consequence orders were placed in England for a steam plough, steel har-

* Based on Memoirs of Department of Agriculture, Department of Agriculture (1954), Madras.

rows, and winnowing machinery and water lifters. An area of 350 acres of land at Saidapet-Madras was taken over by Government to conduct trials with new implements, manures and improved systems of farming. A Superintendent of farm was appointed in 1868 who became Assistant to the Commissioner for Agriculture for general statistics and agricultural work. In 1876, a complete and high class public agricultural college was established at Saidapet which was transferred to Coimbatore in 1907. The main work done during the period was tabulation of village statistics, analysis of famine reports and enquiries on agricultural economics.

The idea of starting agricultural departments in Provinces was first suggested by the Famine Commission of 1880. This resulted in the establishment of agricultural department in various Provinces. In Madras, a separate agricultural department was organised in 1881-82 with a Director of Settlement and Agriculture designated as the Commissioner for Revenue Settlement, Land Records and Agriculture. The need for agricultural research through experimental farms by the department of agriculture was recommended by a Committee which made a comprehensive enquiry into the operation of agricultural department during 1885. Action on the above recommendations was taken only in 1897 when sugarcane crop was affected by red-rot disease in the Godavari area. An Economic Botanist was appointed in 1896 tackle the disease. This was followed by the establishment of experimental farms at Hagari (1901), Kovilpatti (1901), Samalkota (1902), Palur (1905) and Taliparamba, (1905) to undertake agricultural research work on crops like cotton, sugarcane, groundnut and pepper. A whole time Director of Agriculture was appointed in 1906 to work under the Board of Revenue. Besides, Director of Agriculture, the new department also had an Economic Botanist, and Agricultural Chemist and two Deputy Directors (each for North and South Zone). From this small beginning the work of the department expanded as follows :

Agricultural research and experiments: With the appointment of Economic Botanist to undertake plant-breeding work in 1902, the agricultural research in identification of new strains actively started. A separate chemistry section was started in 1909 at Coimbatore to study the problems of soils, manures, and agricultural products. Research work on plant diseases was given to a separate section in 1901 under a Mycologist. The research work was further strengthened by the appointment of an Economic Botanist for rice in 1913 and of a Sugarcane Specialist in 1913. The research work on paddy received greater attention following the opening of paddy breeding station at Coimbatore in 1913. A number of such station for paddy breeding work were also opened during the period between 1922 to 1942.

A Cotton Specialist was appointed in 1921 to follow up the earlier work done during 1909. The cotton breeding station started at Coimbatore in 1922, was concerned with work on Cambodia, Karunganni and Uppam. New strains were evolved and a special variety called 'Uganda was introduced'. This was a cross between Cambodia and South African Uganda and the finest commercial cotton available during 1930s and 1940s. A Cotton Extension Officer and three Assistant Cotton Extension Officers were appointed in 1950 to undertake extension work related to cotton.

A millet breeding station was opened in 1923 in Coimbatore and it was followed by a number of sub-stations. High-yielding types of pearl millet, finger millet and *tenai* were introduced. The research in oil seeds was intensified after 1930s and high-yielding strains of groundnut, and coconuts were evolved. Work on sugarcane was the earliest to start in 1902 to investigate red-rot. A sugarcane breeding station was started in 1912 at Coimbatore and the work expanded by the introduction of hybrid types raised from seed. The canes evolved at the station catered to the needs of a large sugarcane area of India. The Government of India took over control of the breeding station from the year 1925. Separate sections were also created for the work in agricultural engineering (1928), mycology (1910), entomology (1912), livestock (1916) and pulses (1943). From 1902 to 1948, about 58 research stations were opened up for the purpose of studying local farming problems on crops.

Organisation of the department: As reported earlier a full time Director of Agriculture was appointed during 1906. Besides Director, two Deputy Directors were appointed to take up the propaganda work in the two Circles. (North and South). Between 1906 and 1914, the state had only two Circles with headquarters at Bellary and Tiruchirapalli. From 1914 a central Circle was added at St. Thomas Mount. In 1916, the Circles were increased from three to seven with Assistant Directors of Agriculture for every two or three Districts, and a separate Deputy Director, one for livestock and another for 'planting' Districts. Between 1921 and 1928, there were eight Circles but they were reduced to three in 1938 and two in 1941 as a measure of retrenchment. During 1941, the number of Assistant Directors was increased to provide an officer for each District and they were named as District Agricultural

Officer. During the War period the number of Deputy Directors was again raised to eight. The Deputy Directors also had control over the agricultural stations in the District. The organisational set-up during 1940 is given below.

At the headquarters the Director was assisted by two Deputy Directors for guiding the propaganda work in the Districts. The state was divided into eight Circles of 2 to 4 Districts each and each Circle was incharge of a Regional Deputy Director of Agriculture. There were District Agricultural Officers each incharge of a District and the Taluk Agricultural Demonstrators worked under them. Each Taluk had atleast one Agricultural Demonstrator (during 1920s and 1930s there was one Demonstrator for 2 to 3 Taluks) and their main duty was agricultural demonstration work. At District headquarters there were Mycological and Entomological Assistants to attend to the plant protection work. The demonstrators were assisted by fieldmen and demonstration maistries. The fieldmen had educational level upto fourth form (9th standard) and they took care of demonstrations and trial plots. They were employed at the rate of one per every two to four *firkas* (Block). The demonstrators maistries were literate men who actually demonstrated the improved implements, sowing and putting up manure pits. They were employed at the rate of one per *firka*. The Agricultural Demonstrators worked under District Agricultural Officers who worked in close co-operation with Revenue Department and were attached to Collectors since 1941.

Methods of extension or propaganda: The main extension methods followed by the department were : (1) demonstration in ryot's field or in seed and demonstration farms, (2) trial plots or observation plots, (3) exhibition and agricultural shows, (4) publicity through departmental publications, press note and radio talk, and (5) contact with ryots through village and Taluk agricultural associations.

The agricultural department was mainly concerned with the propagation of following improved practices :

(a) Distribution of improved strains of various crops;
(b) Popularisation of green manure crops, rural and urban compost;
(c) Improved method for preservation of farm yard manures;
(d) Distribution of manures like groundnut cake, ammonium sulphate, and phosphate;
(e) Popularisation and distribution of improved farm implements and machineries like improved ploughs, ridger, puddlers, seed

drill, guntakas, oil engines, electric pumps and motors;

(f) Popularisation of plant protection measures using chemicals like DDT, BHC;

(g) Distribution of takkavi loans;

(h) Popularisation of improved agronomic practices like transplanting in paddy, line sowing of cotton and other cultural practices; and

(i) Distribution of vegetable seeds and fruit saplings.

During 1938-39, the department of agriculture of Madras had 126 officers involved in research and teaching (2 in Superior Service, 21 in Provincial Service and 103 in Subordinate Service) and 272 officers involved in District work (2 in Superior Service, 13 in Provincial Service and 257 in Subordinate Service).

The agricultural department was considerably strengthened in terms of staff and budget as a result of Grow More Food Campaign which was initiated in 1943 in order to meet the acute food shortage. Two vernacular journals called *Mezhichelvam* in Tamil and *Padi Pantalu* in Telugu were published to popularise the Grow More Food Scheme and the activities of agricultural department. Distribution of seeds and manures was also intensified. During the middle of 1943 considerable attention was paid for increasing food production. A new two year plan of intensive cultivation was drawn up for the state with a 13-point programme involving a total cost of Rs. 3.5 crores. Part of the cost was met by the Central Government.

The department was expanded after the initiation of Five Year Plan especially with the introduction of National Extension Service in 1952. The Community Development Programme provided a framework for agricultural extension organisation at Block level. Every District was divided into suitable number of Blocks and an Agricultural Extension Officer was appointed in each Block to work under the Block Development Officer as well as the control of District Agricultural Officer. Each Agricultural Extension Officer was supported by a team of field staff consisting of village level workers, fieldmen and maistries.

During 1950s the department of agriculture was also re-organised. Separate major divisions of research, education and extension were formed. Further at the headquarters many posts at the level of Joint Directors or Additional Directors were created to look after the specialised areas like soil conservation, agricultural engineering, plant protection and Intensive Agricultural District Programme. Many posts of Deputy Directors of Agriculture were created to take up the developmental work at regional level. Several changes took place during 1956 as a result of bifurcation of the State, District and Taluks.

A major change in the department occurred during 1971 when the activities of the department pertaining to the basis agricultural research and higher agricultural education was transferred to the newly formed Tamil Nadu Agricultural University, Coimbatore, excepting regional research stations. After mid-1970s many specialised activities of the department with regard to agricultural marketing were separated and vested with newly formed Directorate of Agricultural Marketing. Further, two new departments namely Seed Certification and Horticulture and Plantation Crops were formed during 1979-80. Again during 1980-81, two more departments to deal with the activities of oil seeds and agricultural engineering were formed. In 1981-82, all the regional research stations which were under the control of the department of agriculture were transferred to the Tamil Nadu Agricultural University.

The agricultural extension wing of the department was re-organised by the introduction of Training and Visit System of Extension with the help of World Bank during 1981. As a result considerable changes were made in the extension organisation set up at State, District and Block level along with the creation of many new posts. Entire state has been covered under the Training and Visit System of Extension. At present the department of agriculture deals mainly with cereal crops and its activities are mainly concerned with agricultural extensions, seed production and distribution, soil testing, and quality control of fertilizer and pesticide.

Origin and History of the Department of Agriculture of Madhya Pradesh*

The present Madhya Pradesh was a major part of Central Provinces of British India before Independence and hence the history of department of agriculture of Central Provinces is narrated below. The Central Provinces consisted of present Madhya Pradesh (except Bhopal), a part of Rajasthan and Maharashtra. The total area under cultivation in Central Provinces and Berar during 1881 was 1.19 million with a population of 11.1 million. The population rose to 16.8 million during 1941. The foundation of an agricultural department in the Central Provinces may be considered to date from 1862 when an Agric-

* Based on *Evidences Taken in the Central Provinces and Berar*, Vol. VI (1927). Royal Commission on Agriculture in India, Calcutta: Central Publication Branch.

Horticultural Society was opened in Nagpur. A Cotton Commissioner was appointed in 1866-67 and an agricultural survey of cotton tracts was done as early as 1861.

The agricultural department was established in the Central Province around 1880 following the report of the Famine Enquiry Commission of 1880 which had strongly urged the setting up of permanent departments of agriculture in the provinces. A Commissioner of Agriculture was appointed who also combined the duties of the Commissioner of Settlement. The functions of the Provincial department consisted of collection of statistical information, demonstration of improved techniques of farming, conducting agricultural experiments and multiplication of seeds. The earliest experimental farm was started in Nagpur during 1883 with an area of 90 acres. The Nagpur farm was mainly conducting agronomical and manurial experiments with cotton, sorghum and sesamum. This farm also served to give practical training to the high school students of agriculture of Nagpur. And during 1888 an agricultural institute was established in Nagpur to offer two years course in agriculture. And during 1906 it was offering a three year diploma course.

During 1905, a full time Director of Agriculture was appointed and the Province was divided into suitable number of Circles under the control of Deputy Director of Agriculture. During the period 1905 to 1920, many experimental farms were established for the purpose of evolving improved agronomic practices, selection of improved seed materials multiplication of seeds and testing of improved farm implements. A review of the operations of the agricultural department during the decade from 1905 to 1916 revealed that achievement of the agricultural department has been in the areas like distribution of good seeds of various crops and improved implements through agricultural associations, and propagation of new cultivation practices for crop production. The agricultural research of the department was strengthened during the period 1915 to 1925 by the establishment of various research laboratories and appointment of Agricultural Chemist, Economic Botanist, Mycologist and Agricultural Engineer. The extension and development work of department was also strengthened by the appointment of many officers at Circle and sub-Circle levels. The progress in agricultural research and in extension during 1915 to 1925 is discussed below.

Agricultural Research: A considerable amount of attention was given for plant improvement in the crops like cotton, wheat, jowar and sugarcane. In cotton, attention was paid to improve its quality, length and strength of staple. The early attempts to improve cotton were controlled by Deputy Directors of the Southern Circle. During the War the Botanist went on war service and the work passed from an active to a more or less dormant condition. At the close of War, plant breeding work again became active. A second Botanist was appointed in 1925 who took over the organisation of teaching and research work related to breeding of local wheat and pulse crops. The breeding work on cotton resulted in the separation of improved variety called *roseum* which was spread to about half a million acres in the cotton tract during 1920s, resulting in many crores of rupees benefit to the cultivators. Many improved varieties of crops of sorghum, paddy, wheat and sugarcane were also released.

The satisfactory position of the agricultural department of Central Provinces and Berar in agricultural research during 1920s could be seen from the following evidences of the Director of Agriculture made to the Royal Commission on Agriculture in India (1927).

> "It may be said that perhaps with exception of less millets and some of the less grown oil seeds and pulses the department is in a position at the present day to provide the grower with high yielding and higher quality and better suited varieties of nearly every crop commonly grown."

Extension: The major work of the department related to extension during the period 1905 to 1930 was related to introduction of improved cultivation of major crops based on the work of experimental farms and the expansion of area under improved strains and new crops. Efforts were also made to introduce improved farm implements and machineries. The methods of extension used by the agricultural department were : (1) agricultural shows at farms or fairs, (2) seed and demonstration farms, (3) establishment and management of seed stores and implement depots, (4) demonstration plots, (5) smaller practical demonstration from village to village, and (6) distribution of pamphlets, bulletins and lantern lectures.

The demonstrations laid out in the seed and demonstration farms and in the cultivators' field were mainly used to propogate improved techniques. On an average more than 2400 demonstrations were laid out per year. One of the earlier improved techniques demonstrated was 'transplanting' instead of 'broadcasting' rice. The campaign was organised in 1906-07 and only 20 acres were planted with 'transplanting' During 1918-19 the acreage under transplanted rice rose to 41,825 acres. Another technique demonstrated was line sowing (using country drills) of cotton and sorghum in the place of broadcasting. Use of copper sulphate and copper carbonate as a preventive measure against smut in sorghum also became popular as a result of demonstrations. In sugarcane demonstration of planting of canes in ridges and furrows instead of planting in flat beds was done. A simple furnace for boiling sugarcane juice was also introduced.

Many improved varieties of sesamum, cotton, sugarcane and rice were also introduced. Improved seeds and implements were sold through depots. The introduction of groundnut crop became successful and within a short period it occupied about 30,000 acres.

The organisation of the department: The organisational set up the department during late 1920s and during 1930s was as follows :

At headquarters there was a full time Director of Agriculture. Under the Director there were four Deputy Directors incharge of the four Circles, namely Northern, Western, Southern and Eastern with the headquarters at Jabalpur, Amraoti, Nagpur and Raipur respectively. In addition there was a Deputy Director of Animal Husbandry with headquarters at Nagpur. Among the research and teaching staff, Principal of Agricultural College was stationed at Nagpur. The other research staff included two Economic Botanists, a Mycologist and an Agricultural Engineer. They were all the officers of Imperial Agricultural Services.

Organisation within circle: Each Circle as headed by a Deputy Director of Agriculture. Under each Deputy Directors, there were one or more Provincial Officers who were termed as Extra Assistant Directors of Agriculture. Each Circle had 2-3 Extra Assistant DirectorsAgriculture. They took one or more Districts and were responsible for controlling the demonstration work of agricultural assistant. The Extra Assistant Directors of Agriculture were not concerned with experimental work and under these men there were staff who belonged to the Subordinate Agricultural Service. The Subordinate Agricultural Service was divided into an upper and lower division. People with degree in agriculture were directly recruited to upper division from agricultural college but as a rule promotions were made from lower division to the upper division. These subordinate official staff did entirely propaganda work except those who worked on experimental farms. There were two kinds of farms. The seed and demonstration farm was under the control of Provincial Service Officers. The experimental farms were under the direct control of Deputy Directors.

Each Circle was divided into two sub-Circles, each consisting of 2-3 Districts under the charges of Extra-Assistant Director who was primarily concerned with organisation of demonstrations. He was also incharge of Government seed and demonstration farms which fell within his Circle. The Extra-Assistant Directors were helped by Agricultural Assistants who were also designated as Superintendents and Overseers on Government farms. There were only two Agricultural Assistants per District and one Assistant's charge might include 400-500 villages. Each Agricultural Assistant was helped by two or more Kamdars.

During 1938-39 the department had about 34 officers (2 in Superior Service, 6 in Provincial Service and 26 in Subordinate Service) involved in research and teaching and about 133 officers (3 in Superior Service, 12 in Provincial Service and 118 in Subordinate Service) involved in District work. The department of agriculture was reorganised after the introduction of National Extension Service in 1952. The Community Development Programme provided a framework for agricultural extension organisation at Block level. The department of agriculture has undergone many changes since the formation of new state of Madhya Pradesh which was constituted on 1st November, 1956. It comprises Mahakashal unit of erstwhile Madhya Pradesh state, the whole of the former State of Vindhya Pradesh and Bhopal.

The Directorate of the Department of Agriculture at Bhopal was expanded during 1960s with the creation of many posts of Joint Directors and Deputy Directors to implement various schemes. The organisational set up at Divisional, District and Block levels also expanded by creation of a number of posts. The agricultural research and teaching work of the department was separated and a separate agricultural university namely Jawaharlal Nehru Krishi Vishwa Vidyalaya was formed at Jabalpur. All the research and experimental stations were transferred to the University. During 1966 the post of Block Development Officer was abolished and all the agricultural officers and village level workers were placed under the vertical control of the agricultural department.

The extension wing of the department of agriculture was strengthened with the introduction of Training and Visit System of Extension in the State during 1977-78. As a result many new posts of Subject Matter Specialists, Agricultural Extension Officers and Village Level Workers were created. The linkage of the department with agricultural research was also strengthened. During early 1980s the activities of the department with regard to agricultural marketing as well as horticulture were separated and separate departments for agricultural marketing and horticulture were formed. At present there is also a proposal to separate the activities of agricultural engineering to form a separate department of agricultural engineering.

A Comparative Analysis of History and Origin of the Department of Agriculture of Tamil Nadu and Madhya Pradesh

The comparative analysis of the origin and history of the agricultural departments of Tamil Nadu and Madhya Pradesh shows that Tamil Nadu which is the developed state in agriculture had relatively early and good beginning in agricultural research than that of Madhya Pradesh. While the agricultural research work in Madhya Pradesh started receiving attention only after 1915, the research work in Tamil Nadu received attention as early as 1868 and a strong centre of agriculture was established in Coimbatore. The research work in Central Provinces in the beginning, was centered around Nagpur which became the part of the State of Maharashtra after re-organisation. The entire area of former Bhopal which was a Princely state, did not receive attention in agricultural research or extension work before Independence. The total research efforts in terms of number of agricultural stations and number of scientific workers employed were higher in Tamil Nadu than Madhya Pradesh before Independence. For example during 1937-38 there was one officer (superior and subordinate service) of the department of agriculture for every 80,484 acres of net area sown in Madras Province as against 1,46,934 acres in the case of Central Province. Though the reasons for the above situation could be explained partly due to a relatively large area under intensive cultivation and a higher percentage of area under irrigation in Tamil Nadu even before Independence, the other reasons are due to its (Tamil Nadu) early development of developmental infrastructures. The Madras Province was much more prone for general development since it was easily accessible through sea ports and the educational and other related developmental activities started relatively earlier. Further it should also be noted that the relatively higher population density and the biotic pressure on land in Madras Province should have contributed to the need of paying more attention to agricultural research and extension work. The role of pioneering individuals in Madras Province who took special care in the early establishment of agricultural college as well sugarcane breeding institute should not also be forgotten.

With regard to agricultural extension and developmental work before Independence, both the states had almost a similar organisation set up at State, District and Village level. However, the department of agriculture of Madras Province was relatively stronger even in extension and development work. For example the number of personnel engaged in extension work at District and village level was higher in Madras Province. Apart from the larger area of cultivation that too under irrigation, the other reasons for the relatively stronger position of Madras Province in extension work might be due to its strong research base and early start of developmental work. Further as in the case of research the former Princely state of Bhopal did not pay any significant attention to agricultural extension work before Independence. After Independence the differences in growth of the department of agriculture of both the states could be attributed to the differences in the investment in the agriculture and irrigation and other situational factors.

The above analysis of origin and history of the department of agriculture of the developed as well as the less developed state shows that contextual factors have played an important role in the development of agricultural departments. The formation and growth of agricultural departments in the developed as well as the less developed state has been influenced by different contextual factors. This supports the General Hypothesis, GH-2, which states that there will be differences in the contextual factors of the department of agriculture of the developed state and the less developed state.

Budgeted Expenditure

The budgeted expenditure of the department of agriculture in the states of Tamil Nadu and Madhya Pradesh was analysed in the present investigation. The budgeted expenditure of the departments before Independence and for the recent period (Sixth Plan Period) has been discussed below. The expenditure before Independence refers to the expenditure incurred by Madras Province and Central Province and Berar by the department of agriculture of the respective provinces.

Table 5.1 and Fig. 5.1 shows the pattern of expenditure of agricultural department of Madras Province and Central Provinces during 1937-38. The major heads of expenditure at the time were crops and soils, education and research, animal husbandry and miscellaneous. The total expenditure of department of agriculture of Madras Province was Rs. 22.37 lakh and that of Central Province was Rs. 10.93 lakh. The per capita expenditure on agriculture works out to be 4.53 paise and 6.50 paise respectively for Madras Province and Central Province. During the same period the expenditure of all the state department of agriculture of British India and Indian states was Rs. 179.31 lakh. And the per capita expenditure on agriculture by all the states works out to be 6.02 paise. It is interesting to see that the per capita expenditure of Central Province during 1937-38 was above all India average and greater than that of the Madras Province. The above picture remained the same even during 1944-45 when the expenditure of Madras Province was Rs. 48.11 lakh (per capita expenditure 9.75 paise) and that of Central Province was Rs. 20.66 lakh (per capita expenditure 12.29 paise).

TABLE 5.1: Pattern of expenditure of agricultural departments before Independence (1937-38)

(Rs. in lakh)

Sl. No.	Major items of expenditure	Madras Province	Central Province and Berar	Total for British India and Indian States
1.	Crops and soils	9.02	9.01	–
2.	Education and research	1.49	1.21	–
3.	Animal husbandry	1.39	0.71	–
4.	Miscellaneous	10.47	–	–
	Total	22.37	10.93	179.31

Source: Agriculture and Animal Husbandry in India (1938-39), Imperial Council of Agricultural Research (1941), Manager of Publications, Delhi.

Though the per capita expenditure on agriculture was high in Central Province than that of Madras Province during pre-Independence period it was found that the expenditure per acre of gross cropped area during the above period was high in Madras Province. For the period 1937-38 expenditure of agriculture department worked out to be 6.98 and 4.45 paise per acre of gross cropped area respectively for Madras Province and Central Provinces. And during 1944-45 it rose to 15.29 paise in Madras Province and 8.39 paise in Central Province.

The budgeted expenditure of agricultural departments of Tamil Nadu and Madhya Pradesh after Independence: The expenditure of departments of agriculture of Tamil Nadu and Madhya Pradesh shot up after Independence. During First Five Year Plan period (1951-52 to

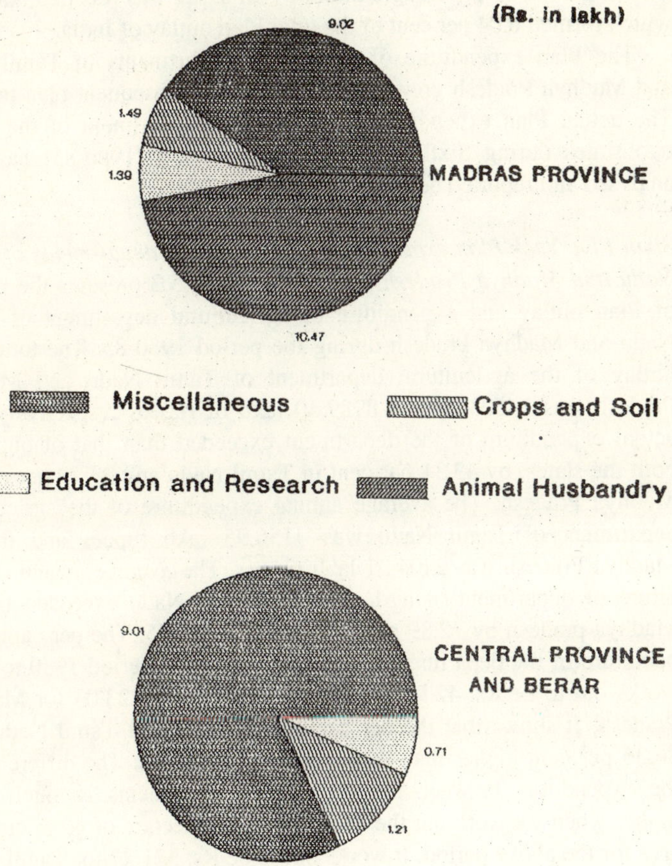

FIG. 5.1: Pattern of expenditure of agriculture departments before independence (1937-38)

1955-56) the outlay of agricultural department (crop husbandry) of Tamil Nadu (then Madras state) was Rs. 1593.7 lakh which was 11.32 per cent of the total Plan outlay of the state. The Plan outlay of agricultural department (crop husbandry) of Madhya Pradesh was Rs. 1005.5 lakh and it formed 23.34 per cent of the total plan outlay of the state. The total Plan outlay on agriculture by all states was Rs. 1,25,000 lakh, which formed 6.04 per cent of the total Plan outlay of India.

The Plan expenditure of agriculture departments of Tamil Nadu and Madhya Pradesh grew in size during the subsequent plan periods. The actual Plan expenditure of agricultural department of the above two states during Sixth Five Year Plan period (1980-85) has been analysed and reported below.

Sixth Five Year Plan expenditure of agricultural departments of Tamil Nadu and Madhya Pradesh (1980-85): Table 5.2 presents the pattern of Plan outlay and expenditure of agricultural department of Tamil Nadu and Madhya Pradesh during the period 1980-85. The total Plan outlay of the agricultural department of Tamil Nadu and Madhya Pradesh was 14503.00 and 8989.40 lakh of rupees respectively. The actual expenditure of the department exceeded than that of outlay in both the states, by 43.31 per cent in Tamil Nadu and 33.73 per cent in Madhya Pradesh. The average annual expenditure of the agricultural department of Tamil Nadu was 4156.95 lakh rupees and that of Madhya Pradesh was 2404.31 lakh rupees. The average annual expenditure of department of agriculture of Tamil Nadu exceeded that of Madhya pradesh by 72.89 per cent during 1980-85. The per capita expenditure of the department of agriculture for the period 1980 to 1985 works out to be Rs. 42.92 for Tamil Nadu and Rs. 23.03 for Madhya Pradesh. It shows that the per capita expenditure of Tamil Nadu was 86.36 per cent higher than that of Madhya Pradesh. The difference in the expenditure between the above two departments is much more higher when we work out the expenditure per hectare of gross cropped area for the above period. It works out to be Rs. 321.27 for Tamil Nadu and Rs. 56.17 for Madhya Pradesh, showing that expenditure of Tamil Nadu is about 472 per cent higher than that of Madhya Pradesh. It should be noted here that Madhya Pradesh has only 14 per cent of its area under irrigation but Tamil Nadu has about 47 per cent of its under irrigation.

It may be worth mentioning here that the per capita outlay of agricultural department (crop husbandry, soil and water conservation and agricultural research education) of Tamil Nadu even during fifth

Five Year Plan period was far higher (Rs. 28.8) than that of Madhya Pradesh (Rs. 8.1).

Table 5.2 shows that the agricultural department of Tamil Nadu spent money on sixteen major schemes and that of Madhya Pradesh spent money on 13 major schemes. In the case of Tamil Nadu the maximum expenditure occurred on multiplication and distribution of seeds (22.07%) followed by commercial crops (19.26%), intensive extension project (T&V) and training of farmers (15.36%), plant protection (14.57%), soil and water conservation (11.94%) and others. The above first five major schemes also consumed about 83 per cent of the total budget.

TABLE 5.2: Plan expenditure of agricultural department of Tamil Nadu and Madhya Pradesh (1980-85)

(Rs. in lakh)

Sl. No.	Name of the Major Scheme		Sixth Five Year Plan outlay	Sixth Five Year Plan expenditure	Average expenditure per year
1	2		3	4	5
1.	Direction and administration	TN	15.00	18.27 (0.09)	3.65
		MP	135.00	49.40 (0.41)	9.88
2.	Intensive Extension Project (T & V system) and training of farmers	TN	2656.25	3192.02 (15.36)	638.40
		MP	3141.50	3321.90 (27.63)	664.38
3.	Multiplication and Distribution of seeds	TN	2022.71	4586.39 (22.07)	917.28
		MP	115.00	123.24 (1.03)	24.65
4.	Agricultural farms	TN	23.05	781.82 (3.76)	156.36
		MP	165.00	218.57 (1.82)	43.71
5.	Manures and fertilizers	TN	257.59	474.82 (2.28)	94.96
		MP	190.00	769.44 (6.40)	153.89
6.	Plant protection	TN	2264.00	3028.73 (14.57)	605.75
		MP	235.00	253.54 (2.11)	50.71

Contd.

TABLE 5.2 *Contd.*

1	2	3	4	5	
7.	Commercial crops	TN	2898.33	4002.12 (1[illegible]26)	800.42
		MP	518.00	830.29 (6.91)	166.06
8.	Horticulture	TN	1239.54	613.67 (2.95)	122.73
		MP	553.90	1543.41 (12.84)	308.68
9.	Agricultural engineering	TN	450.00	300.21 (1.44)	60.04
		MP	230.00	118.77 (0.99)	23.75
10.	Agricultural economics and statistics	TN	—	6.1 (0.03)	1.22
		MP	6.00	1.67 (0.01)	0.33
11.	Storage and warehousing	TN	100.00	121.26 (0.58)	24.25
		MP	—	67.9 (0.56)	13.56
12.	Agricultural marketing	TN	400.00	283.93 (1.37)	56.76
		MP	—	39.1 (0.33)	7.82
13.	Soil and water conservation	TN	1800.00	2480.78 (11.94)	496.16
		MP	3700.00	4684.33 (38.96)	936.87
14.	Dryland development	TN	50.00	51.81 (0.25)	10.36
		MP	—	—	—
15.	Land utilisation	TN	200.38	140.36 (0.86)	28.07
		MP	—	—	—
16.	Other expenditure	TN	126.15	702.44 (3.38)	140.49
		MP	—	—	—
	Total	TN	14503.00	20784.73	4156.95
		MP	8989.40	12021.56	2404.31

TN = Tamil Nadu MP = Madhya Pradesh

Note: Figures in parentheses give percentage

Source: Report of the Five Year Plan (1985-90) and Annual Plan (1986) of Tamil Nadu and Madhya Pradesh.

In the case of Madhya Pradesh, the maximum expenditure occurred on soil and water conservation (38.96%), followed by intensive extension project (T&V) and training of farmers (27.63%), horticulture (12.84%), commercial crops (6.91%), manures and fertilizers (6.40%), and others. The first two major schemes alone consumed about 67.0 per cent of the budget.

It is interesting to see that the state department of agriculture of Tamil Nadu spent a large amount of their budget on multiplication and distribution of seeds (Rs. 4586.39 lakh, 22.07%) and on plant protection (Rs. 3028.73 lakh, 14.57%), whereas in the above areas the department of agriculture of Madhya Pradesh spent only a negligible amount of money (Rs. 123.24 lakh, 1.03% and Rs. 253.54 lakh, 2.11%, respectively). It was also observed that the department of agriculture in Tamil Nadu was quite strong in the supply of seeds to the farmers. Further, the per hectare use of pesticide was also quite high in the state. On the contrary, one of the weak points of the department of agriculture in Madhya Pradesh was on the distribution of improved seeds to the farmers as the seed supply function was almost entirely handled by the co-operatives whose performance was not satisfactory. Further, the state had the lowest consumption of pesticide in India.

It was also observed that the department of agriculture of both the states incurred bulk of the expenditure budgeted for the year during the last quarter of the year. This problem was more acute in the department of agriculture of Madhya Pradesh. The reason for the above state of affair was due to the delay in the financial sanction. As reported by many officials the delay in financial sanction as well as the lack of financial powers of the field level extension officers hindered the efficient functioning of the department of agriculture in both the states.

The above analysis clearly shows that the pattern of expenditure of the department of agriculture of Tamil Nadu and Madhya Pradesh differs from each other before and after the period of Independence in terms of items of expenditure as well as amount of expenditure. Though during pre-Independence period the per capita expenditure of the department of agriculture of Madhya Pradesh was higher than that of Tamil Nadu, the expenditure per acre of gross cropped area was higher in Tamil Nadu during the same period. After Independence the expenditure pattern shows that both per capita as well as per hectare gross cropped area expenditure was much higher in Tamil Nadu than that of Madhya Pradesh. These findings support the General Hypothesis, GH-2, which states that there will be differences in the organisational contextual factors (Budgetted expenditure) of the department of agriculture of the developed state and the less developed state.

6

MACRO-ORGANISATIONAL STRUCTURE

The macro-organisational structure of the departments of agriculture of Tamil Nadu and Madhya Pradesh was studied with respect to the following dimensions:

I. Macro-Organisational Measures

(1) Organisational size
(2) Number of supervisory-level-vertical differentiation
(3) Number of sections and job-titles-horizontal differentiation
(4) Number of field units-spatial differentiation
(5) Technical employees-staff ratio
(6) Ministerial employees-staff ratio
(7) Supporting employees-staff ratio
(8) Administrative Intensity
  (a) Supervisor-staff ratio
  (b) Percentage of supervisory employees above the sub-Divisional unit to the total supervisors
(9) Percentage of posts filled in to the total number of posts sanctioned

II. Organisational Set-up of the Department of Agriculture at State, Divisional, District and Block Level.

First a descriptive account of the macro-organisational measures have been reported for each state separately, which has been followed by a description of organisational set-up of the department at various levels. At the end a comparative analytical account of the above macro-organisational measures for both the states has been given.

DEPARTMENT OF AGRICULTURE, TAMIL NADU

I. Macro-Organisational Structure

Headquarters and Operational Units

The headquarters of the department of agriculture of Tamil Nadu is situated in Madras headed by a Director of Agriculture. It caters to the needs of 4.09 million farm families cultivating a net area of 5.74 million hectares. Out of the net cropped area, about 47 per cent of the area is irrigated. The state has been divided into 17 Divisions each representing a District. The Division is headed by a Joint Director of Agriculture. Every Taluk of the District has been divided into a sub-Division, and there are 99 sub-Divisions each headed by an Assistant Director of Agriculture. The total number of departmental units at Block level is 378. Each agriculture unit at Block has a main agricultural depot to sell improved seeds, pesticides and other inputs. In addition there are 394 sub-agricultural depots. Tamil Nadu has 45 state seed farms each headed by an Assistant Director of Agriculture. There are eight seed processing units, fifteen soil testing laboratories, six fertilizer testing laboratories, nine pesticide testing laboratories and ten parasite breeding centres. There are twelve farmer's training centres in the state.

The total operational units of the state department of agricultural in Tamil Nadu is 1006, spread over in an area of 1.30 lakh square kilometre.

Number of Posts

The total number of posts of the department during 1986 was 14,854 out of which 3,680 (24.77%) belonged to administrative or ministerial staff. And 1,426 (9.60%) posts were of supporting staff (class IV service). The total number of technical posts were 9,778. Among the technical posts only 127 (1.30%) were of class I service (rank of Deputy Director and above) and 315 (3.22%) posts were of class II service (Assistant Director or equivalent rank). The vast majority of the posts, (9,336) which constituted 95.48 per cent of the technical staff belonged to class III service.

The total number of job titles provided in the department was 63. Out of these job titles 15 belonged to administrative staff. Among technical staff there were 48 job titles. The number of important posts of the department and their percentage to the total technical staff during 1985-86 is given in Table 6.1.

Table 6.1 shows that the posts of Assistant Agricultural Officers (village level workers) constituted about 41 per cent of the total technical staff, which was followed by Agricultural Officers (28.7%) Field Demonstration Assistants (14.26%), Deputy Agricultural Officer (4.60%), Assistant Directors of Agriculture (3.22%), Deputy Directors of Agriculture (0.99%), Joint Directors of Agriculture (0.26%) and Additional Directors (0.05%).

TABLE 6.1: Number of technical posts of the department of agriculture of Tamil Nadu

Sl. No.	Name of the post	Number of posts	Percentage of the total technical staff
1.	Director	1	—
2.	Additional Director	5	0.05
3.	Joint Director	25	0.26
4.	Deputy Director (including Agricultural Chemist)	97	0.99
5.	Assistant Director of Agriculture (including Asstt. Agricultural Chemist)	315	3.22
6.	Agricultural Officer	2807	28.70
7.	Deputy Agricultural Officer	450	4.60
8.	Assistant Agricultural Officer	4036	41.28
9.	Field Demonstration Officer	1394	14.26

Number of Posts Filled and Vacancy Position

There were few positions vacant both at the Directorate and at field level. For example during August 1986 only 4 posts of Joint Director (out of 21) and 11 posts (out of 55) of Assistant Agricultural Chemist, and 30 posts (out of 2807) of Agricultural Officers were vacant in the state. At the lower level about 200 posts (out of 4036 sanctioned) of Assistant Agricultural Officers and 190 posts (out of 368 sanctioned) of Assistant Seed Officer were lying vacant. Only few posts under Training and Visit System (T&V) of extension were lying vacant in the state as it can be seen from Table 6.2 which snows staff position of T&V system as on 30.6.1986.

TABLE 6.2: Staff position under training and visit system of extension (Tamil Nadu) as on June, 1986

Sl. No.	Name of the post	Number of posts sanctioned	Number in position	Number vacant
1.	Additional Director (A)	2	2	—
2.	Joint Director (A)	17	17	—
3.	Deputy Director (A)	68	67	1 (1.47)
4.	Assistant Director (A)	103	102	1 (0.97)
5.	Agricultural Officer	1028	1006	22 (2.14)
6.	Assistant Agricultural Officer	4022	3985	37 (0.92)
	Total	5240	5179	61 (1.16)

Note: Figures in the parentheses gives the percentage.

II. Organisational Set-up

State-Level

The department of agriculture of Tamil Nadu is headed by a Director of Agriculture, who executes the government policies regarding the agricultural production through the departmental staff. He is directly responsible to the Secretary of the Department of Agriculture for the functioning of the department. The Agricultural Production Commissioner of the state is the chief co-ordinating authority of various departments and organisations concerned with agricultural production and development.

The headquarters of the department, the Directorate of Agriculture is located at Madras. The Director at headquarters is assisted by four Additional Directors (Fig. 6.1) who are incharge of the major Divisions namely Extension (Training and Visit System), Seeds, Monitoring and Evaluation (Training and Visit system), and Personnel and Planning. The post of Additional Director of Research is kept vacant as per the orders of the Government. The Division of Chemistry and Research is headed each by a Joint Director.

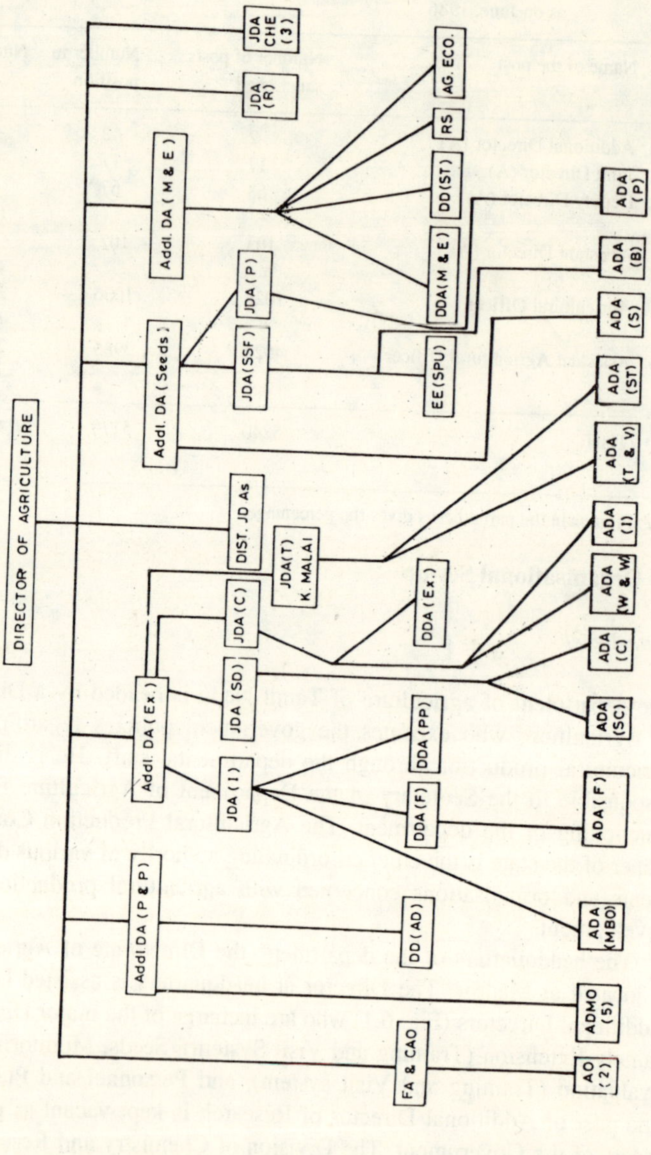

FIG. 6.1: Organisational chart of the department of agriculture of Tamil Nadu (State Level)

Source: Directorate of Agriculture, Department of Agriculture, Madras

Abbreviations

(Figure 6.1)

1. Addl. D.A. (P&P)	Additional Director of Agriculture (Personnel and Planning)
2. Addl. D.A. (Ex)	Additional Director of Agriculture (Extension)
3. Addl. D.A. (M&E)	Additional Director of Agriculture (Monitoring and Evaluation)
4. Addl. D.A. (Seeds)	Additional Director of Agriculture (Seeds)
5. J.D.A. (I)	Joint Director of Agriculture (Inputs)
6. J.D.A. (SD)	Joint Director of Agriculture (Sugarcane Dev.)
7. J.D.A.(C)	Joint Director of Agriculture (Cereals)
8. J.D.A.(SSF)	Joint Director of Agriculture (State Seed Farm)
9. J.D.A.(P)	Joint Director of Agriculture (Pulses)
10. J.D.A.(Res)	Joint Director of Agriculture (Research)
11. J.D.A.(Che)	Joint Director of Agriculture (Chemistry)
12. Dist J.D.A.	District Joint Directors of Agriculture
13. FA & CAO	Financial Adviser and Chief Accounts Officer
14. DD (Admn)	Deputy Director of Agriculture (Administration)
15. DD(F)	Deputy Director of Agriculture (Fertiliser)
16. DD(PP)	Deputy Director of Agriculture (Plant Protection)
17. DD(Ex)	Deputy Director of Agriculture (Extension)
18. DD(M&E)	Deputy Director of Agriculture (Monitoring and Evaluation)
19. DD(ST)	Deputy Director of Agriculture (Statistics)
20. R.S.	Rural Sociologist
21. Agrl. Eco	Agricultural Economist
22. EE(SPU)	Executive Engineer (Seed Processing Unit)
23. AO	Accounts Officer
24. ADMO	Administrative Officer
25. ADO(MBO)	Assistant Director of Agriculture (Management by Objectives)
26. ADA (SC)	Assistant Director of Agriculture (Schools)
27. ADA(C)	Assistant Director of Agriculture (Cotton)
28. ADA(WW)	Assistant Director of Agriculture (Weather and Watch)
29. ADA(I)	Assistant Director of Agriculture (Information)
30. ADA(T&V)	Assistant Director of Agriculture (Training and Visit)
31. ADA(St)	Assistant Director of Agriculture (Statistics)
32. ADA(S)	Assistant Director of Agriculture (Seeds)
33. ADA(P)	Assistant Director of Agriculture (Pulses)
34. JDA(T) K. MALAI	Joint Director of Agriculture (Training) at Kudimian Malai.

There are seven Joint Directors of Agriculture in the Directorate. Two Joint Directors who are incharge of Research and Chemistry sections have been placed directly under the Director and others have been placed under the concerned Additional Director. These Joint

Directors are incharge of sections namely Inputs, Seeds, Cereals, State Seed Farms, Pulses and Sugarcane Development. The entire state has been divided into 19 Divisions on the basis of District and each Division is headed by a Joint Director who works directly under the supervision of the Director.

There are six Deputy Directors of Agriculture who work under Additional and Joint Directors in the area like administration, fertilizer, plant protection, extension, monitoring and evaluation, and statistics. The Directorate has about 10 Assistant Directors and a number of Technical Assistants (in the cadre of Agricultural Officers) who are attached to various sections to assist in carrying out the work of the Directorate.

The technical work of the Directorate has been broadly divided into 15 schemes or sections like Five Year Plan section, Agricultural Chemistry section, Plant Protection section and Farmers Training Centre. The Directorate has twenty administrative establishments like Agricultural Officers establishments, Office Procedure section and Personnel Management section and eight account sections. The total number of technical employees working in the Directorate are about 80 and the number of ministerial employees is about 300.

Divisional Level

Each District of the state has been divided into a Division and the State has a total of 19 Divisions each headed by a Joint Director. The administrative and technical control of all the staff of the agricultural department in the Division is under the control of the Joint Director. Each Division has been divided into sub-Divisions on the basis of Taluk and headed by an Assistant Director of Agriculture.

The Joint Director at his headquarters is assisted by a team of five Deputy Directors (Fig. 6.2). The following are the major sections of the office of the Joint Director, each incharge of a Deputy Director: Training and Visit, System of Extension, Information and Training, Plant Protection, Agronomy and Seeds. The section of Quality Control is headed by an Assistant Director of Agriculture. Each Deputy Director is assisted by an Agricultural Officer.

A typical office of Joint Director has one administrative officer, six superintendents, thirteen Assistants (Upper Division Clerks), seven junior Assistants, four typists, six stenographers, two record clerks, eight drivers and seventeen basic servants (peons). Further it has a provision for an artist, one operator, one tredle man, one compositor, one binder, tow mechanics and two watchmen.

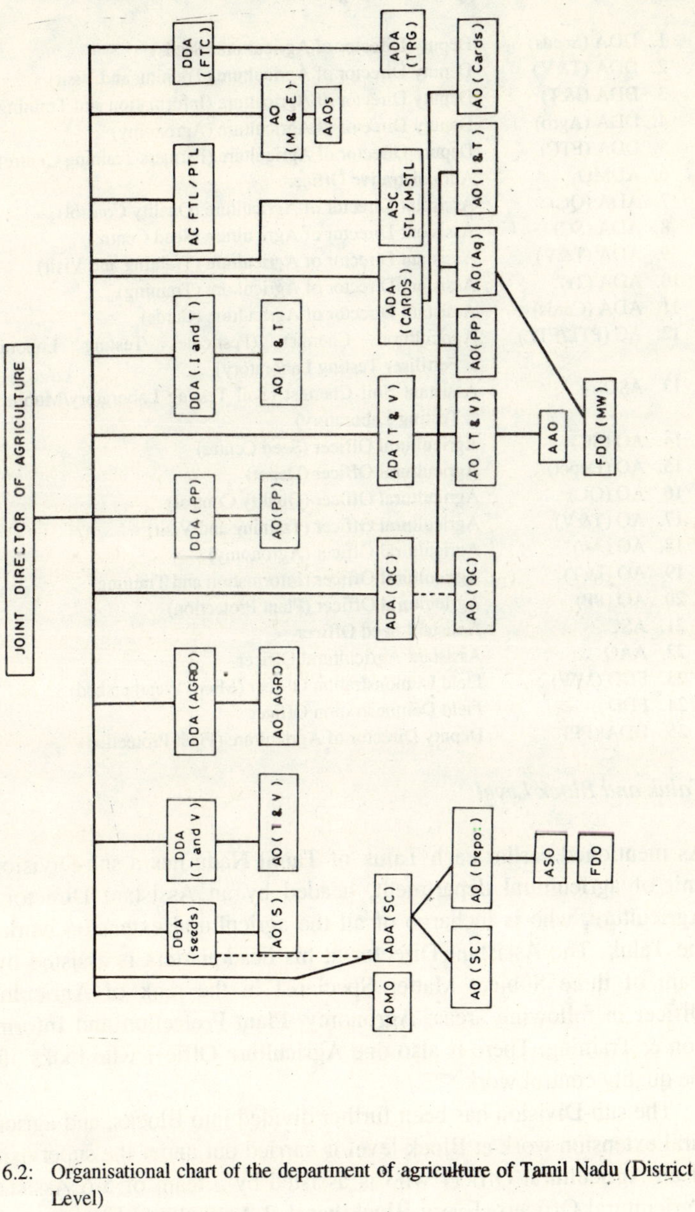

FIG. 6.2: Organisational chart of the department of agriculture of Tamil Nadu (District Level)

Source: Directorate of Agriculture, Department of Agriculture, Madras

Abbreviations

(Figure 6.2)

1. DDA (Seeds)	Deputy Director of Agriculture (Seeds)
2. DDA (T&V)	Deputy Director of Agriculture (Training and Visit)
3. DDA (I&T)	Deputy Director of Agriculture (Information and Training)
4. DDA (Agro)	Deputy Director of Agriculture (Agronomy)
5. DDA (FTC)	Deputy Director of Agriculture (Farmers Training Centre)
6. ADMO	Administrative Officer
7. ADA (QC)	Assistant Director of Agriculture (Quality Control)
8. ADA (SC)	Assistant Director of Agriculture (Seed Centre)
9. ADA (T&V)	Assistant Director of Agriculture (Training and Visit)
10. ADA (Tr)	Assistant Director of Agriculture (Training)
11. ADA (Cards)	Assistant Director of Agriculture (Cards)
12. AC (PTL/FTL)	Agricultural Chemist (Pesticides Testing Laboratory/ Fertiliser Testing Laboratory)
13. ASC	Assistant Soil Chemist (Soil Testing Laboratory/Mobile Soil Testing Laboratory)
14. AO (SC)	Agricultural Officer (Seed Centre)
15. AO (Depot)	Agricultural Officer (Depot)
16. AO (QC)	Agricultural Officer (Quality Control)
17. AO (T&V)	Agricultural Officer (Training and Visit)
18. AO (Ag)	Agricultural Officer (Agronomy)
19. AO (I&T)	Agricultural Officer (Information and Training)
20. AO (PP)	Agricultural Officer (Plant Protection)
21. ASC	Assistant Seed Officer
22. AAO	Assistant Agricultural Officer
23. FDO (MW)	Field Demonstration Officer (Micro Weathershed)
24. FDO	Field Demonstration Officer
25. DDA (PP)	Deputy Director of Agriculture (Plant Protection)

Taluk and Block Level

As mentioned earlier each Taluk of Tamil Nadu has a sub-Divisional unit of agricultural department, headed by an Assistant Director of Agriculture, who is incharge of all the agricultural extension work of the Taluk. The Assistant Director at his headquarters is assisted by a team of three Subject Matter Specialist in the rank of Agriculture Officer in following areas: Agronomy, Plant Protection and Information & Training. There is also one Agriculture Officer who looks after the quality control work.

The sub-Division has been further divided into Blocks, and agricultural extension work at Block level is carried out under the supervision of an Agricultural Officer who is assisted by a team of 5-6 Assistant Agricultural Officers. Every Block has 1-2 Agricultural Depots, under the charge of an Agricultural Officer who works directly under the administrative the technical control of Assistant Director of Agriculture (Seed Centre).

DEPARTMENT OF AGRICULTURE, MADHYA PRADESH

1. Macro-Organisational Structure

Headquarters and Operational Units

The department of agriculture is one of the biggest state government organisations in the state of Madhya Pradesh with a large number of employees and operational units situated throughout the state. It has to cater to the needs of 6.4 million farm families cultivating an area of 19.2 million hectares of land. Out of this net cropped area, only about 14.4 per cent area is irrigated. The department has its headquarters at Bhopal. It has 12 Divisional level units, each headed by a Joint Director, 50 District level units, each headed by a Deputy Director, 110 sub-Divisional units, each headed by a sub-Divisional Agricultural Officer and 450 Block units, each headed by a Senior Agricultural Development Officer. Every District also has a Soil Conservation unit headed by an Assistant Soil Conservation Officer. Further the department has nine Soil Survey units along with their 40 field units or team located in various places. The number of soil testing laboratories in the state is 25. The department also has nine agricultural engineering units located in various places, 101 Government agricultural farms for production of seeds and about 15 other units like farmer's training centres and fertiliser testing laboratory. Thus the department has more than 800 operational units spread over in an area of 4,43,400 square kilometres.

Number of Posts

The total number of posts of the department during 1986 was 30,312. Out of this 3,196 (10.54%) posts belonged to administrative or ministerial staff. And 3,037 (10.01%) posts were of supporting staff belonging to Class IV service. And the total number of technical posts was 24,079. Among the technical posts only 162 (0.67%) belonged to Class I service (Deputy Director and above or it's equivalent post), and 818 (3.40%) posts belonged to Class II service (Assistant Director or equivalent rank) and a thumping majority of posts (23,099) which constituted 95.93 per cent of total technical staff belonged to Class III service (Senior Agricultural Development Officer or its equivalent rank and below). The percentage of Class I and Class II employees in the department of agriculture of Madhya Pradesh is found to be relatively lower when compared to the position in other technical departments and revenue departments. The available data for 1981 show that among the technical staff only 4.51 per cent of employees of department of agriculture belonged to Class I and Class II service. For the same period a higher percentage of technical staff belonged to Class I and Class II services in other departments like animal husbandry (16.10%) public health (14.65%), irrigation (12.67%), public works (8.03%) and revenue (9.66%).

The total number of job titles available in the department was 204. Among these titles 20 belonged to administrative or ministerial staff and 18 belonged to Class IV or supporting staff. Among the technical staff there were 166 job titles. The Class I service had 21 job titles, Class II service had 33 job titles, and Class III had 112 job titles. The sanctioned strength of some of the important categories of the technical staff during 1985-86 is given in Table 6.3. It can be seen from Table 6.3 that the posts of Rural Agricultural Extension Officers constituted about 55 per cent of the total technical staff followed by Agricultural Department Officers (10.30%), Surveyors (9.32%), Senior Agricultural Development Officers (5.82%) and other senior posts.

Number of Posts Filled and Vacancy Position

It was observed that a large number of positions were vacant both at headquarters and at field level for various categories of technical staff. For example, during April 1986 at the headquarters about six post of Additional Director and Joint Director were lying vacant. About 20 posts of Deputy Director were lying vacant.

At field level the extent of vacancy position can be understood from the staff position under Intensive Extension and Research Project of the department during 1986 as reported in Table 6.4.

It can be seen from Table 6.4 that out of the total 9120 posts sanctioned for the project, 3421 posts (37.62%), were lying vacant. A large number of posts of Rural Agricultural Extension Officers, (2106, 38.99%), Agricultural Development Officers (454, 40.97%), Subject Matter Specialist (147, 30.06%) were lying vacant. A considerable number of posts of supervisory level (Deputy Director, Assistant Director) were also not filled in.

TABLE 6.3: Number of technical posts of the department of agriculture of Madhya Pradesh

Sl. No.	Name of the technical post	Number of posts	Percentage of the total technical posts
	Class I service		
1.	Director	1	—
2.	Additional Director	5	0.02
3.	Joint Director	25	0.10
4.	Deputy Director	78	0.32
	Class II service		
5.	Sub-divisional Agricultural Officer (Assistant Director of Agriculture)	564	2.34
	Class III service		
6.	Senior Agricultural Development Officer	1401	5.82
7.	Agricultural Development Officer	2481	10.30
8.	Rural Agricultural Extension Officer (including Rural Horticultural Extension Officer)	13289*	55.19
9.	Agricultural Demonstrator	589	2.45
10.	Surveyors	2249	9.32

* Number of RAEOs under Training and Visit System of Extension = 5402

The reasons for the quite high percentage of posts not being filled up seems to be the administrative problems and inefficiency of recruiting agencies like Public Service Commission and Junior Public Service Commission. For example, it was reported that no recruitment has been done for the post of Agricultural Development Officers though three years has lapsed since the release of advertisement for the recruitment of the posts. Even the adhoc appointment of agricultural graduates has been stopped in recent times. There were many unemployed agricultures graduates in the state and many were willing to take up even the job of Village Level Extension Workers. The paradox is that on one side there are men in search of job, on the other side positions are vacant because of delay on the part of recruitment agencies.

II. Organisational Set-Up

The Department of Agriculture is headed by the Director of Agriculture who is irresponsible for implementation of the Government policies regarding agricultural production in the state. The over all co-ordination of the activities of the various departments and organisations related to agricultural production in the state is done by the Agricultural Production Commissioner of the state.

TABLE 6.4: Staff position under intensive extension and research project (Phase I and Phase II) as on April, 1986 — Madhya Pradesh

Sl. No.	Name of the post	No. of posts sanctioned	No. of posts filled up	No. of posts vacant
1.	Additional Director	1	1	—
2.	Joint Director	3	2	1 (33.23)
3.	Deputy Director	10	5	5 (50.00)
4.	Assistant Director	18	9	9 (50.00)
5.	Sub-Divisional Officer	96	81	15 (15.63)
6.	Subject Matter Specialist (at District and sub-Divisional level)	489	342	147 (30.06)
7.	Agricultural Development Officer	1108	654	454 (40.97)
8.	Rural Agricultural Extension Officer	5402	3296	2106 (38.99)
9.	Rural Sociologist	3	0	3 (100.00)
10.	All other posts (about 35 types of posts)	1990	1299	691 (34.72)
	Total	9120	5689	3431 (37.62)

Note: Figures in parentheses indicate the percentage.

At the headquarters which is located at Bhopal, the Director is assisted by five Additional Directors (Fig. 6.3) who head the five major Divisions namely, Crop Production and Inputs, Pulses and Oilseeds, Dry Farming and Soil Conservation, Intensive Agricultural Extension

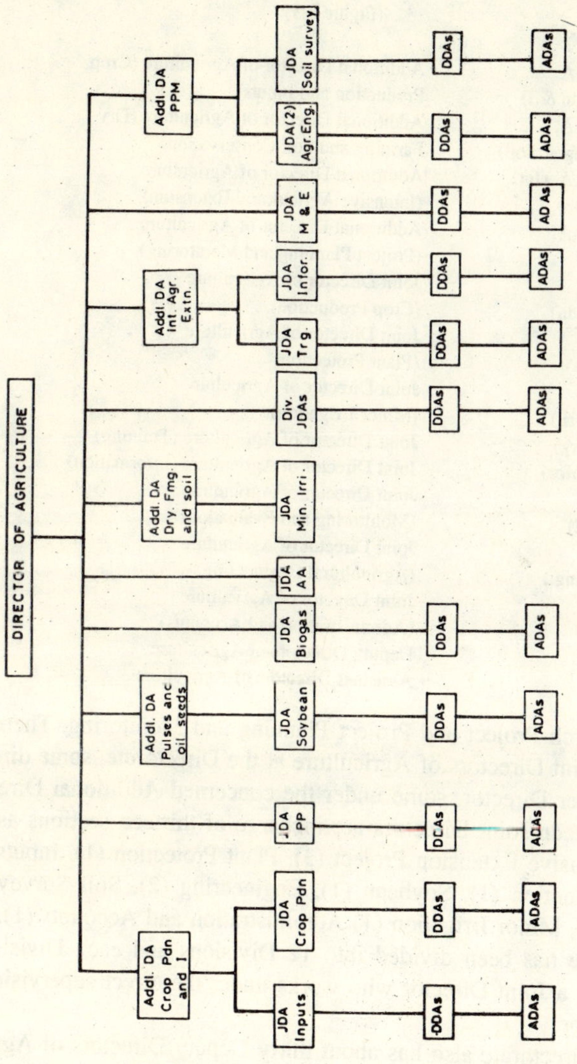

FIG. 6.3: Organisational chart of the department of agriculture of Madhya Pradesh (State Level)

Abbreviations

(Figure 6.3)

1. Addl. D.A. (Crop Pdn & I)	Additional Director of Agriculture (Crop Production and Inputs)
2. Addl. D.A. (Dry Fmg & Soil)	Additional Director of Agriculture (Dry Farming and Soil Conservation)
3. Addl. D.A. (Int. Agr. Extn)	Additional Director of Agriculture (Intensive Agricultural Extension)
4. Addl. D.A. (PPM)	Additional Director of Agriculture (Project Planning and Monitoring)
5. JDA (Crop Pdn)	Joint Director of Agriculture (Crop Production)
6. JDA (PP)	Joint Director of Agriculture (Plant Protection)
7. JDA (Min. Irri.)	Joint Director of Agriculture (Minor Irrigation)
8. JDA (Trg)	Joint Director of Agriculture (Training)
9. JDA (Infor)	Joint Director of Agriculture (Information)
10. JDA (M & E)	Joint Director of Agriculture (Monitoring and Evaluation)
11. JDA (Agr. Eng.)	Joint Director of Agriculture (Agricultural Engineering)
12. JDA (AA)	Joint Director of Agriculture (Administration and Accounts)
13. DDAs	Deputy Directors of Agriculture
14. ADAs	Assistant Directors of Agriculture

and Research Project and Project Planning and Monitoring. There are thirteen Joint Directors of Agriculture at the Directorate, some directly works under Director, some under the concerned Additional Director. These thirteen Joint Directors are incharge of thirteen sections as follows: Intensive Extension Project (3), Plant Protection (1), Inputs (1), Crop Production (1), Soybean (1), Engineering (2), Soil Survey (1), Biogas (1), Minor Irrigation (1) Administration and Accounts (1). The entire state has been divided into 12 Divisions and each Division is headed by a Joint Director who works under the direct supervision of the Director.

The Directorate also has about thirty Deputy Directors of Agriculture incharge of various schemes, some directly work under Director, others under the concerned Joint Director or Additional Director. There are about 40 Assistant Directors of Agriculture, 30 Senior Agricultural Development Officers, besides few Technical Assistants attached to various sections or schemes who help in the functioning of the Directorate.

Divisional Level

The state has been divided into 12 Divisions each headed by a Joint Director. The administrative and technical control of all the staff of agricultural department working in an Agricultural Division is under the control of concerned Joint Director of Agriculture. Each Division comprises of a few Districts and each District is having a Deputy Director of Agriculture and his team of Subject Matter Specialists. The Divisional Soil Conservation Officers as well as the Agricultural Engineer come under the over all control of the Joint Director.

Joint Director at his headquarters is assisted by a team of six Subject Matter Specialists (Fig. 6.4), (in the rank of Assistant Directors) and a few number of Senior Agricultural Development Officers and Agricultural Development Officers. The office of Joint Director has the following six Sections each incharge of an Assistant Director; Land development and Soil conservation, Intensive Extension Project and Training, Tribal sub-plan and 20-Point Programme, Crop Production and Planning, Plant Protection and Inputs, and Establishment and Accounts. Apart from the above technical staff about 40 administrative and supporting staff headed by an Assistant Superintendent assist the Joint Director to carry out his duties. The administrative staff of a typical office of Joint Director consists of one Assistant Superintendent, one Divisional Accountant, three auditors, one accountant, six assistant auditors, five Upper Division Clerks. ten Lower Division Clerks/ Typists, one Daftari, ten Peons, one Mechanic and one Farash.

District and Block Level

There are 45 Districts in the sate of Madhya Pradesh and there are 50 District units of agriculture each headed by a Deputy Director of Agriculture. The Deputy Director is responsible for all the agricultural staff in the District including the work of soil conservation and agricultural engineering. He also guides the work of Government agricultural farms. Each District has been divided into a few sub-Divisions each headed by a sub-Divisional Agricultural Officer. There are 110 sub-Divisions in Madhya Pradesh.

The Deputy Director is assisted by a team of five Subject Matter Specialist (Fig. 6.5), (in the rank of Assistant Directors) in the areas like Agronomy, Special Crops, Training, Plant Protection and Water Management. The Subject Matter Specialists are assisted by Senior Agricultural Development Officers and Agricultural Development Officers.

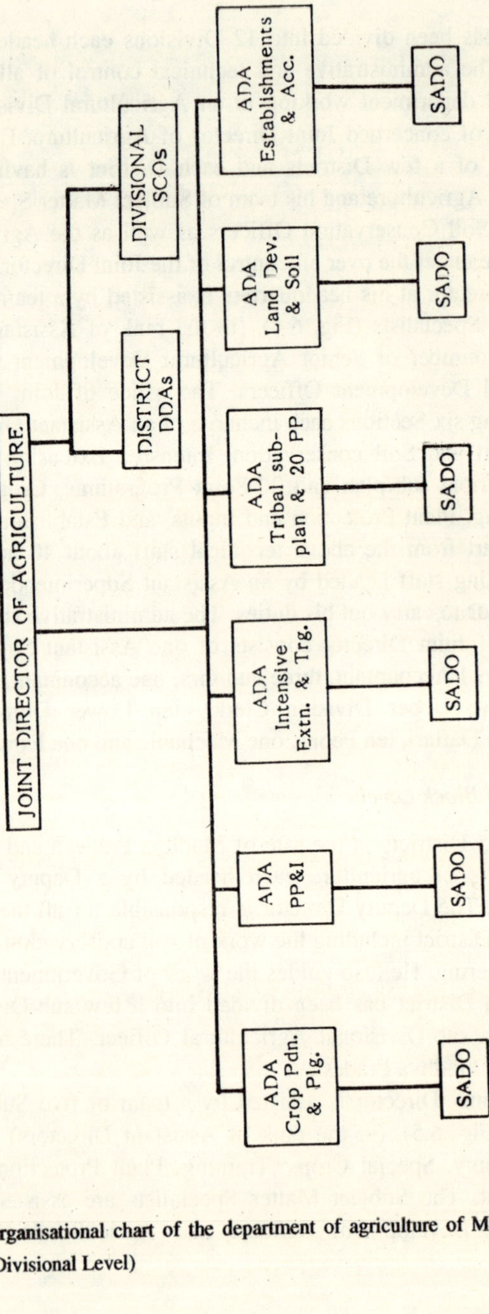

FIG. 6.4: Organisational chart of the department of agriculture of Madhya Pradesh (Divisional Level)

Abbreviations

(Figure 6.4)

1. DDAs	Deputy Director of Agriculture
2. SCOs	Soil Conservation Officers
3. ADA (Crop Pdn & Plg)	Assistant Director of Agriculture (Crop Production and Planning)
4. ADA (PP & I)	Assistant Director of Agriculture (Plant Protection and Inputs)
5. ADA (Intensive Extn & Trg)	Assistant Director of Agriculture (Intensive Extension Project and Training)
6. ADA (Tribal sub-plan & 20 PP)	Assistant Director of Agriculture (Tribal Sub-plan and 20-Point Programme)
7. ADA (Land Dev. & Soil)	Assistant Director of Agriculture (Land Development and Soil Conservation)
8. ADA (Establishment & Acc)	Assistant Director of Agriculture (Establishment and Accounts)
9. SADO	Senior Agricultural Development Officer

There are about 12-15 administrative and supporting staff headed by a Head-clerk who assist the Deputy Director in carrying out his duties. A normal sanctioned set up the office of the Deputy Director has the following ministerial staff: one Head-clerk, one Stenographer, one Accountant, three Upper Division Clerks, six Lower Division Clerks, one Driver, four Peons and one Watchman.

The District is divided into 2-3 sub-Divisions. Each sub-Division is headed by a Sub-Divisional Agricultural Officer in the rank of Assistant Director. Each Assistant Director is assisted by three Subject Matter Specialists in the rank of Senior Agricultural Development Officer, five clerical staff and three peons. The sub-Divisional Agricultural Officer directly supervises the agricultural staff at the Block level. At Block level, there is one Senior Agricultural Development Officer who is assisted by 4-6 Agricultural Development Officers. Each Agricultural Development Officer is assisted by six Rural Agricultural Extension Officers. The Village Level Workers have been designated as Rural Agricultural Extension Officers.

A Comparative Analysis of Macro-Organisational Structural Configuration of the Department of Agriculture of Tamil Nadu and Madhya Pradesh

Table 6.5 shows various measures of macro-organisational structural configuration of the department of agriculture of Tamil Nadu and Madhya Pradesh, a comparative account of which has been narrated below:

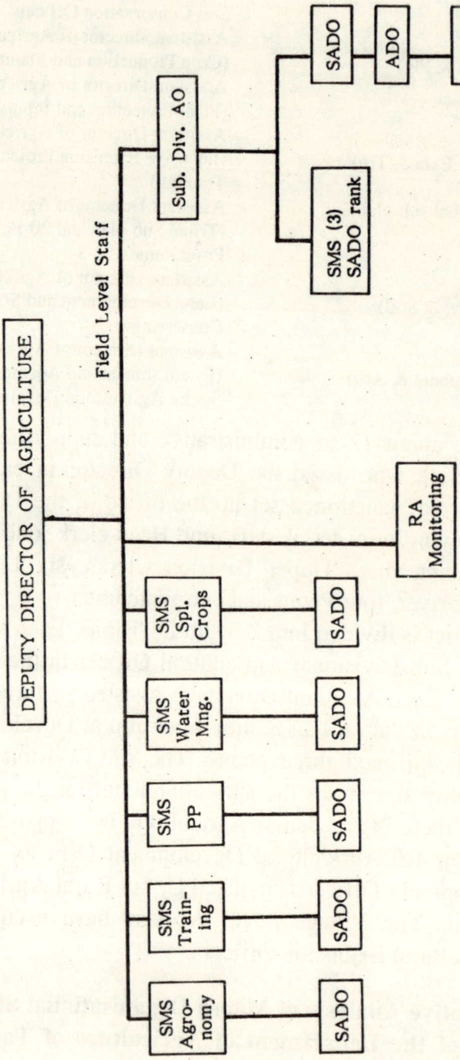

FIG 6.5: Organisational chart of the department of agriculture of Madhya Pradesh (District Level)

Abbreviations

(Figure 6.5)

1. Sub. Div. AO	Sub-Divisional Agricultural Officer
2. SMS (PP)	Subject Matter Specialist (Plant Protection)
3. SMS (Water Mng)	Subject Matter Specialist (Water Management)
4. SMS (Spl Crops)	Subject Matter Specialist (Special Crop)
5. SADO	Senior Agricultural Development Officer
6. ADO	Agricultural Development Officer
7. RA	Research Assistant
8. RAEO	Rural Agricultural Extension Officer.

Organisational size: It is seen from Table 6.5 that the size of the department of Madhya Pradesh measured in terms of its total employees was 2.03 times higher than that of Tamil Nadu. It is worth to note here that the department of agriculture of Madhya Pradesh caters to the need of a higher population which is 1.56 times higher than that of Tamil Nadu and a larger net cultivated area (2.34 times higher than that of Tamil Nadu). The number of technical staff of the department per 1000 cultivators work out to be 2.35 in Madhya Pradesh and 1.79 in Tamil Nadu. The lower technical staff cultivator ratio, in Tamil Nadu is due to the fact that its department of agriculture has been bifurcated into many other departments like Horticulture, Oil-seeds, Agricultural Engineering and the present department of agriculture mainly consists of extension wing, seed wing and the quality control wing. But the department of agriculture of Madhya Pradesh at the time of investigation consisted among other wings, agricultural engineering wing, oilseeds wing and a part of horticultural wing.

The VLW (engaged in crop husbandry work)–staff ratio did not vary between the two states though it was slightly higher in Tamil Nadu than that of Madhya Pradesh. The VLW-staff ratio worked out to be 0.698 for Tamil Nadu and 0.062 for Madhya Pradesh. Thus it is clear that the intensity of field level extension functionary does not vary between the department of agriculture of developed state and less developed state. However, the number of farm families per VLW under Training and Visit System of Extension was relatively lower in Tamil Nadu (1016) than that of Madhya Pradesh (1187). This may be due to the fact that Tamil Nadu has a higher percentage of area under irrigation where a comparatively higher VLW-farmer ratio is prescribed.

TABLE 6.5: Measures of macro-organisational structural configuration of agricultural department

Sl. No.	Measures of organisational structural configuration	Department of agriculture	
		Tamil Nadu	Madhya Pradesh
1.	Organisational size (Total number of employees)	14884	30312
2.	Number of supervisory levels	6	7
3.	Number of major sections	7	13
4.	Number of job titles	48	166
5.	Number of field units	1006	871
6.	Technical employees-staff ratio	1.915	3.860
7.	Ministerial employees-staff ratio	0.328	0.119
8.	Supporting employees-staff ratio	0.106	0.113
9.	Administrative intensity		
	(a) Supervisor-staff ratio	0.047	0.042
	(b) Percentage of supervisory employees above the sub-Divisional unit to the total supervisors	28.70	16.20
10	Percentage of posts filled into the total number of posts sanctioned	98.83	62.37

Number of supervisory level: It indicates the vertical differentiation of the organisation. The number of supervisory level in the department of Agriculture, Tamil Nadu was six (AO — Asstt DA — DDA — JDA — ADA — DA) and that of department of Agriculture of Madhya Pradesh was seven (ADO — SADO — SDAO — DDA — JDA — ADA — DA). And it should be noted that unlike in Madhya Pradesh the Deputy Directors of Tamil Nadu did not have independent office but were stationed at the office of Joint Directors and functioned mainly as Subject Matter Specialists. Thus it can be seen that vertical differentiation of the department of agriculture of Madhya Pradesh was higher than that of Tamil Nadu. It should be pointed out here that though vertical differentiation may provide increased orderlines and review of organisational process, but by providing more levels between the top and bottom of the organisation, it may cause so much distortion of information that control may be lost rather than enhanced (Pfeffer, 1978).

Number of job titles and sections: The number of job titles and the number of sections indicates the horizontal differential of the organisation. It can be seen from Table 6.5 that the department of agriculture of

Madhya Pradesh had a higher level of horizontal differentiation than that of Tamil Nadu. The department of agriculture of Madhya Pradesh had 166 technical job titles and 13 major sections, while that of Tamil Nadu had 48 job titles and seven major sections. The reasons for less number of job titles and major sections in the department of agriculture of Tamil Nadu is due to the fact that it has been bifurcated into various departments or Directorates, as pointed out in the previous chapter.

Spatial differentiation: It refers to the total number of geographical sites or field units of the organisation. It is interesting to note that though the size of the department of agriculture of Madhya Pradesh is 2.03 times greater than that of Tamil Nadu the number of its geographical units or field units (871) is about 13 per cent lower than that of Tamil Nadu (1006). This is due to the fact that the department of agriculture of Tamil Nadu has one or two depots for supply of farm inputs at every Block. This may also suggest that the field unit of department of agriculture of Tamil Nadu is smaller in size and less differentiated internally than that of field units of Madhya Pradesh, but has a greater decentralisation and supervisory intensity at the service-delivery level, which may increase the visibility of the organisation and job satisfaction of the employees (Van de ven and Ferry, 1980). The increase in spatial differentiation may result in increased administrative overhead cost in each field unit, but it results in increased accessibility and adaptability of an organisation to its clients.

Technical employees-staff ratio: It refers to the ratio between the technical employees and the total non-technical employees of the organisation. Table 6.5 shows that the technical employees staff ratio was much higher in the department of agriculture of Madhya Pradesh (3.860) than that of Tamil Nadu (1.915). This indicates that the intensity of technical employees is higher in Madhya Pradesh. The reason for this may be due to its greater organisational size and larger field or geographical units.

Ministerial employees-staff ratio: It refers to the ratio between ministerial employees and the total non-ministerial employees of the organisation. The ministerial employees-staff ratio of the department of agriculture of Tamil Nadu (0.328) was much higher than that of Madhya Pradesh (0.119). This clearly indicates that intensity of ministerial or administrative employees is higher in Tamil Nadu. It was also observed during investigation that while Tamil Nadu did not face any problem in carrying out administrative work of the department, Madhya Pradesh was facing problems with administrative work, due to shortage of number of employees or inefficient functioning. Though there are no research data available on the desirable ministerial employees-staff ratio especially for the developmental departments, it can be safely said here that the ratio of ministerial employees-staff in the department of agriculture of Madhya Pradesh needs suitable modification so as to improve the functioning of its ministerial wing.

Supporting employees-staff ratio: It refers to the ratio between supporting employees (Class IV) and the total non-supporting employees of the organisation. Table 6.5 shows that the supporting employees-staff ratio of the department of agriculture of Tamil Nadu (0.106) was almost the same as that of Madhya Pradesh (0.113). This indicates that the intensity of supporting staff is almost the same in both the department.

Administrative intensity: The administrative intensity of the department of agriculture was studied for technical staff in terms of the following : (a) Average supervisor-staff ratio, (b) percentage of supervisors (b) above the sub-Divisional unit (above Assistant Director level) to the total supervisors.

Table 6.5 shows that the supervisory-staff ratio of the department of agriculture of Tamil Nadu (0.047) and Madhya Pradesh (0.042) did not significantly differ from each other and was comparatively low. This shows the shape of the pyramid of organisational set up of department of agriculture of both the state is flat, having less administrative intensity. This is also one of the main reasons for lack of promotional opportunities for junior level staff in both the states. However, the percentage of supervisors above the sub-Divisional unit (above Assistant Director level) to the total supervisors was high in Tamil Nadu (28.7%) than that of Madhya Pradesh (16.2%). This reflects that the proportion of middle and upper level supervisors to the total supervisors of the department in Tamil Nadu was higher than that of Madhya Pradesh. Thus it is seen that over all administrative intensity of both the departments of agriculture is low, though the department of agriculture of Tamil Nadu has a higher proportion of middle and upper level of supervisors.

Percentage of post filled up to the total number of posts sanctioned: The number of posts sanctioned does not give the correct picture about the employees strength in the organisation many a time, due to the problem of not filling up the posts or vacancy positions in time. The percentage of post filled up to the total number of posts sanctioned indicates the percentage of employees in position at a particular point of period. The percentage of post filled up to the total number of posts sanctioned was worked out for the staff position under Intensive Extension and Research Project for which the complete data was available. Table 6.5 shows that the percentage of the posts filled up was much higher in the department of agriculture of Tamil Nadu (98.83%) than that of Madhya Pradesh (62.37%). The agricultural department of Madhya Pradesh had 37.63 per cent of its posts as vacant. This indicates that percentage of employees in position was much higher in Tamil Nadu than Madhya Pradesh.

The above analysis clearly shows that the department of agriculture of Tamil Nadu and Madhya Pradesh differ from each other with respect to many measures of macro-organisational structural configuration. The department of agriculture of Madhya Pradesh was relatively high in organisational size, horizontal differentiation, vertical differentiation, and technical employees-staff ratio. The department of agriculture of Tamil Nadu had relatively higher spatial differentiation, ministerial employees-staff ratio and a higher percentage of posts filled in to the total number of posts sanctioned. Though the supervisor staff-ratio of both the departments was almost the same, the department of agriculture of Tamil Nadu had a higher proportion of middle and upper level supervisors than that of Madhya Pradesh.

One of the major differences between the department of agriculture of Tamil Nadu and Madhya Pradesh was seen in the bifurcation of the department into many related departments or Directorates like Agricultural Engineering, Horticulture, Oilseeds etc. While the department of agriculture of Tamil Nadu was highly bifurcated, its counterpart in Madhya Pradesh was less bifurcated. It was interesting to note the present department of agriculture in Tamil Nadu was concerned mainly with cereals and its activities related to extension (Training and Visit System), seed production, quality control and soil testing.

In the organisational set up of the department of agriculture it was interesting to see the Divisional level offices (comprising 2-3 Districts) which were present in Madhya Pradesh were missing in Tamil Nadu though every District had been divided into a Division headed each by a Joint Director. As a result, unlike in Madhya Pradesh, the Deputy Directors of Agriculture of Tamil Nadu did not have independent offices but were stationed at the office of Joint Directors and functioned mainly as Subject Matter Specialists.

At the Block level it was seen that there were two cadres of extension officers in Madhya Pradesh one known as Agricultural Development Officer (A.D.O.) and another Senior Agricultural Development Officer (S.A.D.O.) who was the head of the extension set up at Block level. But in Tamil Nadu there was only one cadre of extension officer known as Agricultural Officer (A.O.).

The above analysis clearly points out that the department of agriculture of Tamil Nadu and Madhya Pradesh considerably differ from each other on the aspects of macro-organisational structural configuration. This supports the General Hypothesis, GH-4, which states that there will be differences in the macro-organisational structural configuration of the department of agriculture of the developed state and the less developed state.

7

RECRUITMENT, SALARY AND PROMOTIONAL POLICIES

The personnel management practices of the department of agriculture of Tamil Nadu and Madhya Pradesh was studied in the present investigation, with regard to recruitment, salary and promotion and the same have been discussed below (Table 7.1 and 7.2).

Village Extension Worker

It was observed that both in the case of Tamil Nadu and Madhya Pradesh, the lowest level extension functionary, i.e. the village level extension worker was recruited directly. However, in Tamil Nadu about 10 per cent of village extension workers was recruited through promotion of the Field Demonstration Officers (FDO) who were generally posted in seed depots of the Department. In Tamil Nadu, the village level extension worker was designated as Assistant Agricultural Officer (AAO) and was recruited by the state department of agriculture. In Madhya Pradesh, the village level extension worker was designated as Rural Agricultural Extension Officer (RAEO) and was recruited through Junior Public Service Commission Board. In Tamil Nadu the minimum qualification for direct recruitment was diploma in agriculture after high school pass. However, it was observed in Madhya Pradesh, that the candidates with pass in Higher Secondary School Examination or who were graduates in any discipline were selected and posted as RAEOs after six months training. There are few RAEO's training centres in Madhya Pradesh. It has been decided recently to recruit only the graduates for post of RAEOs in Madhya Pradesh. The initial monthly salary (pay + dearness allowance) of AAO in Tamil Nadu was Rs. 766 and that of RAEO in Madhya Pradesh was Rs. 965.

TABLE 7.1: Recruitment, promotion and salary of agricultural personnel (Tamil Nadu)

Sl. No.	Cadre	Recruitment method: Direct %	Direct Qualifications	Through promotion %	Through promotion Qualifications	Salary*
1.	Assistant Agricultural Officer (VEW)	90	Diploma in Agriculture	10	10th pass + 10-13 years experience	766
2.	Deputy Agricultural Officer	—	—	100	Promotion of A.A.O.	1123
3.	Agricultural Extension Officer (AEO)	100	B.Sc. (Ag.)	—	—	1243
4.	Assistant Director of Agriculture	—	—	100	B.Sc. (Ag.) + Seniority	1423
5.	Deputy Director of Agriculture	—	—	100	B.Sc. (Ag.) + Seniority	1930
6.	Joint Director of Agriculture	—	—	100	B.Sc. (Ag.) + Seniority	2479
7.	Additional Director	—	—	100	B.Sc. (Ag.) + Seniority	2930

* Salary = Basic pay at the start of the grade + dearness allowance per month as on August 1986.

TABLE 7.2: Recruitment, promotion and salary of agricultural personnel (Madhya Pradesh)

Sl. No.	Cadre	Recruitment method				Salary*
		Direct		Through promotion		
		%	Qualifications	%	Qualifications	
1.	Rural Agricultural Extension Officer (V.E.W)	100	10th class/ Graduate	—	—	965
2.	Agricultural Development Officer (ADO)	60	B.Sc. (Ag.)	40	Promotion of R.A.E.O. + Seniority	1125
3.	Senior Agricultural Development Officer	10	M.Sc. (Ag.) or B.Tech.	90	Seniority	1316
4.	Assistant Director of Agriculture	50	M.Sc. (Ag.) or B. Tech.	50	Seniority	1416
5.	Deputy Director of Agriculture	25	B.Sc. (Ag.) with 5 years experience or M.Sc. (Ag.)	75	Seniority	1920
6.	Joint Director	—	—	100	Seniority	2667
7.	Additional Director	—	—	100	Seniority	2920

* Salary = Basic pay at the start of the grade + dearness allowance per month as on April, 1986.

Agricultural Extension Officer

At Block level in Tamil Nadu there was only one cadre of Agriculture Extension Officer, known as Agricultural Officer (AO) though there was also a cadre known as Deputy Agricultural Officer (Dy. AO) who is a promotee from the cadre of village level extension worker (i.e. Assistant Agricultural Officer). All AOs were directly recruited and had a minimum qualification of B.Sc.(Ag.) The initial monthly salary of Dy. AO was Rs. 1123. The monthly salary of AO was Rs. 1243 and the pay scale and status of AO was equivalent to that of Assistant Veterinary Surgeon or Junior Engineer. All AOs had gazetted status of the State Government. In the case of Madhya Pradesh at Block level there were two cadres of extension officers, one is known as Agricultural Development Officer (ADO) and another Senior Agricultural Development Officer (SADO) who was the Head of the agricultural extension set up at Block level. About 60 per cent of the posts of ADO was filled up through direct recruitment of candidates with a minimum qualification of B.Sc. (Ag.) and the rest of 40 per cent was filled up by promotion of RAEOs. In the case of SADOs about 10 per cent of the posts was filled up through direct recruitment of candidates with M.Sc. (Ag.) and the rest 90 per cent through promotion of ADOs. The starting monthly salary of ADO was Rs. 1125 and that of SADO Rs. 1316.

It was observed that neither ADO nor SADO had gazetted status of the State Government and their status and pay scale was much lower than that of their counterparts in other departments. For example, all the Veterinary and Engineering graduates had initial entry as Assistant Veterinary Surgeon or Assistant Engineer and were placed in the scale of Rs. 1000-1920, but most of the graduate of agriculture had initial entry as ADOs with a pay scale of Rs. 740-1180. Even the next promotional cadre of ADOs (i.e. SADOs) had a pay scale of Rs. 900-1450. In fact, only the pay scale of Assistant Director of Agriculture was equivalent to the pay scale of Assistant Veterinary Surgeon or Assistant Engineer. It is interesting to see that once the pay scale for the initial entry of agricultural graduate was even higher than that of veterinary or engineering graduates. Over the years the pay scale of agricultural graduates has been reduced. This has resulted in great resentment among the field level agricultural extension officers. And they resorted to a state level strike during 1986 to press their demand for the parity of pay scale with veterinary graduates. But the strike was unsuccessful as it failed to evoke any favourable decision from the State Government.

Senior Level Posts

All senior level posts in Tamil Nadu, like Assistant Director, and above were filled up by promotion. There was no provision for direct recruitment though a few years ago about 10 per cent of the posts of District Agricultural Officers was directly recruited among the candidates having M.Sc. (Ag.) qualification. This procedure has been stopped because of the strong protest from the Association of Agricultural Extension Officers of the department. In contrast, in the case of Madhya Pradesh, about 50 per cent of the posts of Assistant Director was directly recruited from among the candidates having M.Sc. (Ag.) or B.Tech. Further, 25 per cent of the posts of Deputy Director was directly recruited from among the candidates having a qualification of B.Sc. (Ag.) with five years experience or M.Sc. (Ag.). The posts like Joint Director or Additional Director were filled up entirely through promotion. The direct recruitment to the senior level posts was done through Public Service Commission of Madhya Pradesh. The salary of the senior level officers of both the states was more or less same except the salary of the Joint Director was higher in Madhya Pradesh than that of Tamil Nadu.

Promotion

It was observed that there was a stagnation of the staff of department of agriculture in both the states without sufficient promotional opportunities. The problem was more acute especially among the lower level extension functionaries like VEW or AEO There were many incidences of village level workers (AAO) retiring from the service without getting even a single promotion. The channel of promotion of village level worker in Tamil Nadu is as follows:

A.A.O. ----------------> Dy. A.O.
18-20 years (Deputy Agricultural Officer)

The village level worker who is known as Assistant Agricultural Officer (AAO) can hardly expect one promotion in 18-20 years to the post of Deputy Agricultural Officer which is slightly lower than the Agricultural Extension Officer (AEO) cadre. The non-agricultural graduates can not aspire to get the post of Agricultural Officer in Tamil Nadu. But there is a provision of selection grade for A.A.O. after 10 years.

The channel of promotion of Agricultural Extension Officers (agricultural graduates) in Tamil Nadu is as follows:

A.O. ----> Asstt. D.A. ----> D.D.A. ----> J.D.A. ----> Addl. D.A.
18-20 years | 10-12 years | 6-7 years | 6-7 years

The agricultural graduate who newly joins as Agricultural Officer can hope to get promotion up to Additional Director of Agriculture, but at present the time lag of promotion points out that he can at the maximum become Deputy Director of Agriculture.

The time lag of promotion of Agricultural Officer to the next cadre of Assistant Director of Agriculture is about 18-20 years. But there is a provision for selection grade for Agricultural Officer after a service of 10 years. The time lag for promotion of Assistant Director of Agriculture (Asstt. D.A.), Deputy Director of Agriculture (DDA) and Joint Director of Agriculture (JDA) to the next cadre is 10-12 years, 6-7 years, and 6-7 years respectively.

The channel of promotion of village level worker (non-agricultural graduate) in Madhya Pradesh is as follows:

RAEO ------------> ADO ------------> SADO
20-22 years | 15 years

The village level extension workers who is designated as Rural Agricultural Extension Officer (RAEO) can hope to become Agricultural Development Officer (ADO), Senior Agricultural Development Officer (SADO) and also Assistant Director of Agriculture. But the time lag for the promotion to the next cadre for RAEO is 19-22 years and that of ADO is 15 years and that of SADO is 14 years A small percentage of RAEOs is also sent for B.Sc. (Ag.) course to enable them to improve their competence and promotional opportunities But as in the case of Tamil Nadu, there is no provision of selection of grade post for RAEOs.

The channel of promotion of Agricultural Extension Officer (agricultural graduates) of Madhya Pradesh is as follows:

ADO ----> SADO ----> Astt. D.A. ----> DDA ----> JDA ----> Addl. D.A.
15 years | 14 years | 14 years | 6 years | 6 years

At present an agricultural graduate who joins as ADO can hope to become Additional Director of Agriculture but the time lag promotion is quite high that he can hardly get about 2 or at the maximum 3 promotions in his entire service with a possibility of reaching to the level of Deputy Director of Agriculture. The time lag for promotion of ADO to the next cadre of SADO is about 15 years. And there is no system of selection grade post for ADOs. The time lag for the promotion of Senior Agricultural Development Officer (SADO), Assistant Director of Agriculture (Asstt. D.A.), Deputy Director of Agriculture (DDA) and Joint Director of Agriculture to the next cadre is 14 years, 14 years, 6 years, and 6 years respectively.

The above analysis of selected aspects of personnel management practices of the department of agriculture of Tamil Nadu and Madhya Pradesh shows that they considerably differ from one another in its policies on recruitment, promotion and amount of salary. The initial amount of salary for village level extension worker was relatively high in Madhya Pradesh though the initial amount of salary of agricultural extension officer was relatively high in Tamil Nadu. There was no significant difference in the salary of senior level officers. While the official status and pay scale of Block level agricultural extension officers of Tamil Nadu was at par with Veterinary Assistant Surgeon or Junior Engineer, the status of Block level extension officers in Madhya Pradesh was far lower than that of their counterparts in other technical departments. Though the time lag for promotion of village level worker was more or less the same in both the states, the time lag for the promotion of Block level extension officer was relatively high in Madhya Pradesh. It would take about 29 years for a fresh agricultural graduate in Madhya Pradesh to become Assistant Director of Agriculture through normal channel of promotion while it took only about 18 years for the above promotion in Tamil Nadu. Further in Tamil Nadu there was a provision of selection grade for village level extension workers as well as agricultural extension officers. However, it should be noted that while Tamil Nadu had no provision for direct recruitment of officers beyond and level of Agricultural Extension Officers, such provision existed in Madhya Pradesh up to the level of Deputy Director of Agriculture. Thus it can be concluded that the personnel management practices like recruitment, promotion and salary of department of agriculture of Tamil Nadu considerably differed from that of Madhya Pradesh. On the whole the personnel management practices of department of agriculture of Tamil Nadu are relatively more favourable towards its employees than that of the department of agriculture of

Madhya Pradesh. The above findings support the General Hypothesis, GH-3, which states that there will be difference in the selected personnel management practices like recruitment, promotion and salary of the department of agriculture of the developed state and the less developed state.

8

UNIT STRUCTURE AND PROCESSES

The findings with regard to the various dimensions of unit structure and processes of the department of agriculture of the developed state and the less developed state are discussed in this chapter. The findings have been presented in the following order : (1) Socio-personal characteristics of the employees, (2) Job-design characteristics, (3) Unit structural characteristics, (4) Intra-unit communication flows, (5) Work group functioning, (6) Supervisory behaviour, (7) Time use, (8) Job technology, (9) Pay administration, and (10) Creativity climate.

In this chapter and in the following chapters the use of the names of the developed state as well as the less developed state selected for the study has been purposively avoided in order to avoid the repetition of names of the states. It is needless to say that the developed state refers to Tamil Nadu and the less developed sate refers to Madhya Pradesh.

It should be mentioned here that the respondents of the department of agriculture of the developed state and that the less developed state have been classified into High, Medium and Low categories on various scores of unit structural characteristics and processes based on the standard deviation and mean of the total respondents. The respondents who had the score falling within the range of standard deviation plus or minus mean were classified as having medium score. Those who had a score above standard deviation plus the mean score fell into high score category and those who had less score than the standard deviation minus the mean score fell into low score category.

Socio-Personal Characteristics of the Employees

Table 8.1 presents the mean, standard deviation and frequency distribution of the employees of the department of agriculture in the developed state and the less developed state, on the selected socio-personal characteristics. The employees of the developed state had an average age of 38.45 years with a family size of 4.82 members and had 15.67 years of service in the organisation. And the employees of the less developed state had an average age of 39.83 years with a family size of 5.6 members and had 17.55 years of total service in the organisation. The average salary of the employees of the developed state was Rs. 1,635 per month and that of the less developed state was Rs. 1,542.

Both the groups of employees differed significantly from each other with only regard to family size, though the employees of the developed state were relatively lower in age, having less number of total service and more average monthly salary.

It can be seen from Table 8.1 that most of the employees of both the departments of agriculture fell into medium category with regard to the age, family size, number of years of service and salary per month. A considerable number of the employees of the developed state was lower in age (18.67%), total years of service (26.67%) and salary per month (22.67%) than that of employees of less developed state (10.00, 15.33 and 14.66 per cent, respectively).

It is clear from the above results that the employees of agricultural department of the developed state and the less developed state did not differ significantly from each other on socio-personal characteristics except with regard to average family size. Thus it can be concluded that the employees of the department of agriculture of the developed state and the less developed state do not differ from each other on socio-personal characteristics. These findings do not support the General Hypothesis (GH-5) which states that there will be difference in the socio-personal characteristics of the employees of the department of agriculture of the developed state and the less developed state.

Inter-Correlations Among Socio-Personal Characteristics

Since there were a number of variables in each of the organisational dimensions under study, it was found necessary to work out the inter-correlations among them to know about the inter-relationship.

Table 8.2 presents the inter-correlations among the socio-personal variables. It shows a high degree of relationship among the variables, for both the groups. The results of correlations indicate that salary increases as the age, family size, and total years of service increase for both the groups. One notable difference is that family size increases with age and total years of service with regard to employees of the less developed state and no such relationship is found among the employees of department of agriculture in the developed state.

TABLE 8.1: Mean scores, standard deviation and frequency distribution of respondents on socio-personal dimensions

Dimensions		Range	Mean	S.D.	High score		Medium score		Low score		Value of 't'
					No.	%	No.	%	No.	%	
1. Age	DS	18-56	38.45	8.83	35	23.33	87	58.00	28	18.67	
	LDS	20-57	39.83	8.12	31	20.66	104	69.33	15	10.00	-1.41^{NS}
2. Family size	DS	2-10	4.82	1.65	23	15.33	89	59.33	38	25.33	
	LDS	2-12	5.63	2.23	39	26.00	82	54.66	29	19.33	-3.53^{*}
3. Total years of service	DS	1-32	15.67	8.87	25	16.67	85	56.67	40	26.67	
	LDS	1-34	17.55	8.22	31	20.66	96	64.00	23	15.33	-1.90^{NS}
4. Salary per month	DS	882-3547	1635.51	643.90	32	21.33	84	56.00	34	22.67	
	LDS	987-3518	1542.00	612.22	18	12.00	110	73.33	22	14.66	1.29^{NS}

DS= Developed State
* Significant at 0.05 level of probability

LDS = Less Developed State
NS = Not Significant.

TABLE 8.2: Inter-correlations among the socio-personal variables

Dimensions	DS				LDS			
	1	2	3	4	1	2	3	4
1. Age	1.00				1.00			
2. Family size	0.02	1.00			0.19*	1.00		
3. Total years of service	0.19*	0.05	1.00		0.82**	0.23**	1.00	
4. Salary per month	0.50*	0.22*	0.48**	1.00	0.50**	0.59**	0.71**	1.00

DS= Developed state
** Significant at 0.01 level of probability

LDS = Less Developed State
* Significant at 0.05 level of probability.

Job-Design Characteristics

The data in Table 8.3 reveal that the employees of the agriculture department of the developed state significantly differed from that of their counterparts in the less developed state with regard to seven out of eight characteristics of job-design. The employees of the developed state had significantly higher job exercise, job standardisation, job authority, job accountability, job feed back, expectation of rewards and lower job pressure. Both the groups did not differ with regard to expectation of sanctions.

It can be observed from the data that most of the employees of both the groups had medium score in the job-design characteristics. One glaring difference between the employees of the developed state and that of the less developed state was their distribution in the high and low score. A relatively higher percentage of the developed state employees were found to have high score on all positive dimensions of job-design characteristics (varying from 25.33 to 34.67%, average 27.11%) than that of the less developed state employees whose distribution on high score was very low (2 to 8%, average 5.11%). The picture was of contrast with regard to the distribution of employees on low score on the above dimensions. A higher percentage of the employees of the less developed state had received low score (28 to 42.67%, average 32.11%) when compared to the employees of the developed state (2 to 8%, average 4.44%). A relatively high percentage of the employees of the developed state (38% and 12.67%) had low score on negative items like job pressure and expectation of sanctions than that of the employees of the less developed state (8% and 6.67% respectively).

The above results clearly indicate that the job-design of the employees of the department of agriculture in the developed state was relatively better than that of their counterparts in the less developed state. The overall expertise of the employees of the developed state in terms of their academic qualifications, training and reading habit was greater than their counterparts in the less developed state. Through Training and Visit System was in operation in both the states to impart training to its employees, it was observed during the field visit that training was done with great care, faithfulness and as per the schedule in the developed state than in the less developed state. This would have improved the expertise of the staff of the developed state. In the less developed state it was observed that the training sessions not only started very late but also concentrated mainly on routine administrative

TABLE 8.3: Mean, standard deviations and frequency distribution of respondents on job-design characteristics score

Dimensions		Range	Mean	S.D.	High score		Medium score		Low score		Value of 't'
					No.	%	No.	%	No.	%	
1. Job expertise	DS	3-15	7.78	1.94	41	27.33	103	68.67	6	4.00	7.82**
	LDS	3-15	6.04	1.90	10	6.67	98	65.33	42	28.00	
2. Job standardi-	DS	6-30	18.62	2.93	38	25.33	109	72.67	3	2.00	
sation	LDS	6-30	16.63	4.94	13	8.67	92	61.33	45	30.00	4.25**
3. Job authority	DS	4-20	14.81	2.84	52	34.67	94	62.67	4	2.66	
	LDS	4-20	13.39	4.94	6	4.00	97	64.67	47	31.33	3.05**
4. Job pressure	DS	5-25	12.66	2.27	14	9.33	79	52.67	57	38.00	
	LDS	5-25	15.58	2.92	65	43.33	73	48.66	12	8.00	—8.89**
5. Job accountability	DS	6-30	23.47	2.94	48	32.00	99	66.00	3	2.00	
	LDS	6-30	19.73	4.03	8	5.33	93	62.00	49	32.66	9.17**
6. Job feed back	DS	6-30	18.07	3.10	50	33.33	88	58.67	12	8.00	
	LDS	6-30	15.15	3.19	9	6.00	77	51.33	64	42.67	8.03**
7. Expectation	DS	3-15	6.14	2.00	15	10.00	123	82.00	12	8.00	
of rewards	LDS	3-15	5.16	5.48	0	0.00	108	72.00	42	28.00	2.19*
8. Expectation of	DS	2-10	6.13	2.09	28	18.67	103	68.66	19	12.67	
sanctions	LDS	2-10	6.28	2.47	21	14.00	119	79.33	10	6.67	0.567[NS]

DS= Developed State LDS = Less Developed State

** Significant at 0.01 level of probability * Significant at 0.05 level of probability

NS = Not Significant.

matters. The successful training sessions and the regular monthly or review meetings at District and Divisional level might have also contributed to a higher job feed back in the developed state. The employees of the developed state had higher degree of job standardisation, job authority and job accountability. The field observation and discussion with the employees also confirmed the above findings. It was found that departmental procedures and rules with regard to job or work were well established in the department of agriculture of the developed state and most of the employees were aware about them. The job chart especially under Training and Visit System was available at each unit at District Block level. But in contrast though there were rules and procedures with regard to job or work, most of the employees of less developed state were not aware of them or readily available. It was even difficult to get a job chart not only at field level but also at the headquarters of the department. The greater degree of job authority and job accountability perceived by the employees of the developed state speak of greater delegation of authority among the employees.

The employees of the less developed state had higher job pressure and expectation of sanctions. They were also found to have higher work load and were not able to control the pace of the work. This might be due to the lack of job standardisation or job authority. Further it was also found that many of the employees had additional charges of other posts as a number of posts were vacant. And this situation was found from headquarters to the Block level. Under such pressure of work, the employees could only expect sanctions for not completing their task in time. But in contrast the employees of the developed state had significantly higher score on expectation of rewards. It was interesting to observe that the department of agriculture in the developed state had instituted a number of non-monetary reward in the form of certificate or medals for the best performance in various fields like farm information and production. Further the employees of the developed state reported to have received more verbal appreciation for completing the work in time and for best performance. Such a system of reward was absent in the less developed state. Both the groups of employees did not have monetary rewards or rewards through promotion.

From the above results and discussion, it can be concluded that the overall job-design characteristics of the employees of the developed state was better and significantly differed from that of employees of the less developed state. This supports the General Hypothesis, GH-6, which states that there will be differences in the job design characteristics of the employees of the department of agriculture of the developed state and the less developed state.

Inter-Correlations Among Job-Design Variables

Table 8.4 shows that the results of inter-correlation among the job-design variables for both the groups of employees. It shows a considerable amount of relationship among the variables. The results for both the groups of employees indicate that as the job expertise in terms of formal education and training increases there is significant increase in the standardisation of work procedures and job feed back. The increase in job standardisation also results in increase of job authority in terms of making job related decisions. As job authority increases there is a significant increase in the job accountability. The increase in job authority results in significant increase in the expectation of rewards among the employees.

The results of correlation in Table 8.4 for the developed state employees indicate the following significant relationship which is not seen among the employees of the less developed state : (1) job expertise is positively correlated with job authority, (2) job standardisation is positively correlated with job feed back, and (3) job feed back is positively correlated with expectation of rewards.

The correlations further shows that among the employees of the less developed state an increase in job standardisation results in significant increase in job accountability and expectation of rewards. Further an increase in job accountability results in significant increase in job pressure. The increase in job pressure results in significant decrease in the expectation of rewards but in the expectation of sanctions. The increase in job feed back results in significant decrease in the expectation of sanctions.

Unit Structural Characteristics

The data in Table 8.5 indicate that the employees of the department of agriculture of the developed state significantly differed from the employees of the less developed state with regard to three out of seven unit structural characteristics. The employees of the developed state had significantly higher score on unit characteristics such as unit standardisation, unit employee authority and unit incentives. The above two groups of employees did not differ significantly with regard to characteristics like role interchangeability, supervisory authority, external authority and unit sanctions.

TABLE 8.4: Inter-correlation among the job-design variables

Dimensions	DS 1	2	3	4	5	6	7	8	LDS 1	2	3	4	5	6	7	8
1. Job expertise	1.00								1.00							
2. Job standardisation	0.30**	1.00							0.18*	1.00						
3. Job authority	0.31**	0.27**	1.00						0.09	0.79**	1.00					
4. Job pressure	—0.13	0.02	0.03	1.00					—0.02	0.03	0.12	1.00				
5. Job accountability	0.01	—0.10	0.16*	—0.14	1.00				—0.10	0.52**	0.62**	0.19*	1.00			
6. Job feed back	0.25**	0.35**	0.11	—0.10	0.03	1.00			0.24**	0.06	—0.09	—0.13	—0.09	1.00		
7. Expectation of rewards	—0.09	0.02	0.16*	0.00	0.12	0.17*	1.00		—0.08	0.06	0.18*	—0.28**	0.06	0.03	1.00	
8. Expectation of sanctions	—0.03	—0.02	—0.15	0.14	—0.02	0.01	—0.09	1.00	—0.12	0.51**	—0.07	0.24**	0.11	—0.16*	—0.15	1.00

DS = Developed State
** Significant at 0.01 level of probability

LDS = Less Developed State
* Significant at 0.05 level of probability.

It can be seen from the data of Table 8.5 that most of the employees of both states had medium score on unit structural characteristics. But it is their distribution of on high and low score which distinguished one group from another. A higher percentage of the developed state employees fell under high score on characteristics like role interchangeability (12.33%), unit standardisation (32.00%), unit employee's authority (28.67%), supervisory authority (25.33%) and unit incentives (26.00%) than that of employees of less developed state (6.67, 12.67, 8, 19.33, 14.67 per cent, respectively).

In contrast, the percentage of the employees of the developed state who fell under low score for the above characteristics was quite low when compared to the percentage of the employees of the less developed state who fell under this category.

Though both the categories of employees did not differ significantly with regard to their perception on external authority, a relatively higher percentage of the employees of the developed state (DS = 17.33%, LDS = 4.67%) perceived high external influence of Divisional or State level functionaries on unit functioning with regard to its task, performance and rules and procedures to be followed. Likewise, a relatively higher percentage of the employees of the less developed state (DS = 18.67%, LDS = 25.33%) perceived higher unit sanctions.

The results indicate that the units of agricultural department of the developed state at Block, District and Divisional level have significantly higher standardised operational procedures and performance expectations than that of units of agricultural department of the less developed state. The findings are also in conformity with the field observations, which revealed that lack of coordination, control and improper evaluation of unit activities were some of the major problems of unit administration of the less developed state. These problems were essentially due to lack of standard operating procedures, rules and expectation of performance. Though there were manuals on rules and procedures to be followed they were not readily available or inadequate or impractical on many occasions. The employees were also found to ignore the operational procedures and rules. Further the above problems were aggravated due to the multiplicity of programmes under operation. A typical Office of Deputy Director of Agriculture in a District, in the less developed state was operating as man as 30 programmes, and about 35-40 reports were to be sent to the Divisional office regularly every month.

TABLE 8.5: – Mean, standard deviations and frequency distribution of respondent on unit structural characteristics score

Dimensions		Range	Mean	S.D.	High score		Medium score		Low score		Value of 't'
					No.	%	No.	%	No.	%	
1. Role interchange-ability	DS	3-15	11.72	3.92	32	21.33	107	71.33	11	7.33	
	LDS	3-15	10.98	3.50	10	6.67	115	76.66	25	16.67	1.72^{NS}
2. Unit standar-disation	DS	5-25	17.09	2.95	48	32.00	101	67.33	1	0.67	
	LDS	5-25	13.87	3.75	19	12.67	79	52.67	52	34.67	11.16**
3. Unit employee authority	DS	2-10	5.38	1.71	43	28.67	78	52.00	29	19.33	
	LDS	2-10	4.54	2.05	12	8.00	87	58.00	51	34.00	3.86**
4. Supervisory authority	DS	2-10	8.13	2.31	38	25.33	91	60.66	21	14.00	
	LDS	2-10	7.69	3.37	29	19.33	94	62.66	27	18.00	1.32^{NS}
5. External authority	DS	3-15	12.67	1.81	26	17.33	119	79.33	5	3.33	
	LDS	3-15	12.31	5.91	7	4.67	121	80.67	22	14.67	0.71^{NS}
6. Unit incentives	DS	2-10	5.97	1.63	39	26.00	80	53.33	31	20.67	
	LDS	2-10	5.06	2.50	22	14.67	87	58.00	41	27.33	3.72*
7. Unit sanctions	DS	2-10	4.85	1.74	28	18.67	98	65.33	24	16.00	
	LDS	2-10	3.99	4.76	38	25.33	94	62.67	18	12.00	0.20^{NS}

DS = Developed state
** Significant at 0.01 level of probability
LDS = Less Developed State
* Significant at 0.05 level of probability
NS = Not significant.

With regard to distribution of unit authority, it was found that perceived unit employees authority i.e. the employees say in unit tasks, performance criteria, rules, policies and procedure was significantly higher among employees of the developed state than that of the less developed state. Thus it is clear that units of agricultural departments of the developed state were relatively more decentralised than that of units of less developed state. Further the employees of the developed state perceived a significantly higher incentives than that of employees of the less developed state. As it was already pointed out earlier the agriculture department of the developed state had instituted a number of non-monetary incentives at Divisional, District and Block level in the form of appreciation certificates. Further the closeness of supervision, effective training session and review meetings might have contributed to incentives in the form of verbal appreciation and encouragement to individuals and groups.

As discussed above the units of the department of agriculture of the developed state significantly differed from that of the less developed state with respect to three most important unit characteristics like unit standardisation, unit employee authority and unit incentives though both the departments did not differ with respect to unit characteristics like role interchangeability, supervisory authority, external authority and unit sanctions. And it can be concluded that the department of agriculture of the developed state differed from department of the less developed state with regard to their unit characteristics. This supports the General Hypothesis, GH-7, which state that there will be differences in the unit characteristics of the above two departments of agriculture.

Inter-Correlations Among the Unit Structural Characteristics

Table 8.6 shows the inter-correlations among the unit structural characteristics for the employees of both the states. In the case of both the groups of employees the correlations indicate that as the unit standardisation increases there is significant increase in the unit employees authority but a decrease in the supervisory authority. Further as the unit employees authority increases there is significant decrease in the supervisory authority, but an increase in the unit incentives. The correlations also indicate that as the supervisory authority increases there is significant increase in the unit sanctions. An increase in the unit incentives results in significant decrease in unit sanctions.

TABLE 8.6: Inter-correlations among the dimensions of unit structural characteristics

Dimensions	DS							LDS						
	1	2	3	4	5	6	7	1	2	3	4	5	6	7
Role interchange-ability	1.00							1.00						
Unit standardi-sation	0.05	1.00						0.02	1.00					
Unit employees authority	0.07	0.16*	1.00					0.03	0.18*	1.00				
Supervisory authority	–0.01	–0.18*	–0.19*	1.00				0.07	–0.16**	–0.52*	1.00			
External authority	–0.06	0.02	0.01	0.10	1.00			0.06	–0.03	0.06	0.12	1.00		
Unit incentives	0.02	–0.01	0.27**	–0.08	–0.12	1.00		0.06	0.28**	0.60**	–0.07	–0.18*	1.00	
Unit sanctions	0.02	–0.04	–0.13	0.28**	0.12	–0.30**	1.00	0.06	–0.02	0.12	0.48**	0.20*	–0.40**	1.00

DS = Developed State
** Significant at 0.01 level of probability

LDS = Less Developed State
* Significant at 0.05 level of probability.

Correlations in Table 8.6 shows that among the employees of the less developed state (which is not found among the employees of developed state) an increase in unit standardisation results in increase in unit incentive. The correlations also indicate that an increase in external authority results in decrease in unit incentives but an increase in unit sanction.

Intra-Unit Communication or Information Flows

The exchange of information is the very essence of an organisation. The accomplishment of work depends upon the communication between people in each organisation at various levels. Communication is thus a social process of the broad relevance in the functioning of an organisation.

Information flows in the present study refers to work related messages sent among unit personnel through three modes of communication : written communication, personal communication and group communication. Table 8.7 presents the mean scores, standard deviation and frequency distribution of the respondents of the above three kinds of information flows. It can be seen from the results that the employees of the developed state did not significantly differ from the employees of the less developed state with regard to their score in written communication, personal communication and group communication.

Table 8.7 further reveals that a majority of the employees of both the states had medium score on the three modes of information flows. The percentage of the developed state employees who fell under high score on the three modes of information flows ranged from 16.67 to 23.33 (average 20.44%) and in the case of the employees of the less developed state the above range was 10 to 12.67 per cent (average 12.22%). The percentage of the developed state employees who fell under low score on all dimensions of information flows ranged from 1.33 to 11.33 (average 7.77%) and in the case of the employees of the less developed state the above range was 6.67 to 18.67 per cent with an average of 13.33 per cent.

The results indicate that the amount of information flows in terms of written, personal and group communication in various units of the department of agriculture of the developed state and the less developed state did not vary significantly from one another. Thus it can be concluded that the picture of unit information flows in the above two departments is same. These findings do not support the General Hypothesis, GH-8, which states that there will be differences in the information or communication flows of the department of agriculture of the developed state and the less developed state.

TABLE 8.7: Mean scores, standard deviation and frequency distribution of respondents on unit communication or information flows

Dimensions of communication flows		Range	Mean	S.D.	High score		Medium score		Low score		Value of 't'
					No.	%	No.	%	No.	%	
1. Written communication	DS	4-20	10.87	3.77	35	23.33	98	65.33	17	11.33	
	LDS	4-20	10.51	4.30	19	12.67	103	68.67	28	18.67	0.77^{NS}
2. Personal communication	DS	3-15	9.78	3.78	32	21.33	102	68.00	16	10.67	
	LDS	3-15	9.40	3.21	21	14.00	107	71.33	22	14.67	1.05^{NS}
3. Group communication	DS	4-20	10.20	2.87	25	16.67	123	82.00	2	1.33	
	LDS	4-20	9.90	7.28	15	10.00	125	83.33	10	6.67	0.47^{NS}

DS = Developed State LDS = Less Developed State
NS = Not Significant

Inter-Correlations Among the Dimensions of Communication

Table 8.8 shows that the results of inter-correlations among the modes of information flows. It is clear from the results that the written communication, personal communication and group communication are significantly related to each other.

TABLE 8.8: Inter-correlations among the communication or information flows

Information flow dimensions	DS			LDS		
	1	2	3	1	2	3
Written communication	1.00			1.00		
Personal communication	0.58**	1.00		0.17*	1.00	
Group communication	0.23**	0.20*	1.00	0.49**	0.19*	1.00

** Significant at 0.01 level of probability DS = Developed State
* Significant at 0.05 level of probability LDS = Less Developed State.

The written communication was positively and significantly associated with personal communication and group communication. This means the employees who had higher amount of written communica tion was positively and significantly associated with group communication. The employees who were higher in personal communication also had a high amount of group communication. Likewise the personal communication was positively and significantly associated with group communication. The employees who had higher amount of personal communication also had higher amount of group communication. It can be concluded from the results of correlations that written communication, personal communication and group communication in an organisation is interrelated with each other and the increase in one aspect of information flows results in the significant increase in the other aspects.

Work Group Functioning

Work groups are very important features of the work environment. The work place phenomena can be understood from the functioning of work group. Table 8.9 present the range, mean scores, standard deviation and frequency distribution of respondents on work group functioning dimensions. It can be seen from the results that the employees of department of agriculture in the developed state significantly differed from their counterparts in the less developed state with regard to three out of five dimensions of work group functioning. The employees of the developed state had significantly higher group clarity, group cohesiveness and homogeneity. Both the groups of employees did not differ significantly with regard to group fragmentation and group process.

The results further reveal that a majority of both the groups of employees had medium score with respect to work group functioning. A considerably higher percentage of the employees of the developed state was found to have high score on group clarity (27.67%), group cohesiveness (37.33%), and group homogeneity (32.00%) than that of the employees of the less developed state (4.0, 17.33 and 18 per cent respectively). But in contrast a relatively higher percentage of the employees of the less developed state had low score on group clarity (24.67%), group cohesiveness (30.67%) and group homogeneity (24.00%) than that of employees of the developed state (3.33, 9.33 and 6.00 per cent respectively).

The overall picture of work group functioning shows that the work group phenomenon of the department of agriculture in the developed state is better than that of the department of agriculture in the less developed state in terms of its higher group clarity, group cohesiveness and homogeneity. It is interesting to see that both the groups of employees did not differ significantly with respect to internal fragmentation and open group process. This means that the work group phenomena with regard to sharing their feelings, decision-making as well as occasional bickering among the members did not differ significantly among the employees of both groups. However, as the findings indicated the employees of the less developed state compared to their counterparts in the developed state lacked clear idea about the group's goals. Further they also lacked group cohesiveness. The work group as reported by many respondents tended to discuss their own personal problems and other matters rather than work related matter. This reflects the poor leadership at various levels in the department of agriculture of the less developed state. The employees of the developed state had relatively higher score on group homogeneity than that of the employees of the less developed state. This means that the department of agriculture in the less developed state has members with widely

TABLE 8.9: Mean scores, standard deviation and frequency distribution of respondents on work group dimension

Dimensions		Range	Mean	S.D.	High score		Medium score		Low score		Value of 't'
					No.	%	No.	%	No.	%	
1. Group goal clarity	DS	2-14	10.89	2.29	40	26.67	105	70.00	5	3.33	
	LDS	2-14	9.58	2.72	6	4.00	107	71.33	37	24.67	4.49**
2. Internal fragmentation	DS	3-21	8.11	3.15	23	15.33	89	59.33	38	25.33	
	LDS	3-21	8.00	3.11	17	11.33	86	57.33	47	31.33	0.30^NS
3. Open group process	DS	4-28	17.19	4.06	30	20.00	102	68.00	18	12.00	
	LDS	4-28	17.18	5.64	32	21.33	87	58.00	31	20.67	0.01^NS
4. Group cohesiveness	DS	2-14	10.31	2.29	56	37.33	80	53.33	14	9.33	
	LDS	2-14	8.75	2.16	26	17.33	78	52.00	46	30.67	6.07**
5. Group homogeneity	DS	2-14	9.19	2.61	48	32.00	93	62.00	9	6.00	
	LDS	2-14	7.98	2.45	27	18.00	87	58.00	36	24.00	4.13**

DS= Developed State
** Significant at 0.01 level of probability

LDS = Less Developed State
NS = Not Significant

varying backgrounds, skills and abilities when compared to the department of agriculture of the developed state. The reason might be due to the recruitment policy of the department of agriculture of the less developed state. As the work group has a strong effect on organisation member's job experience, attention has to be given to improve the work group functioning in the department of agricultural of the less developed state especially in the areas of cohesiveness and clarity of group's work goal. The cohesion and goal orientation of the group as a social work unit is often a critical determinants of the group's effects on organisational members.

From the above findings it can be concluded that the work group functioning of the department of agriculture of the developed state significantly differed from that of the less developed state. The work group functioning in terms of group clarity and cohesiveness was better in the department of agriculture of the developed state. This supports the General Hypothesis, GH-9, which states that there will be difference in the work group functioning of the department of agriculture of the developed state and the less developed state.

Inter-Correlations Among the Work Group Functioning Dimensions

Table 8.10 shows the inter-correlations among the work group variables for both the groups of employees. It shows a considerable amount of relationship among the variables. The results of correlation for both group of employees indicate that as the group goal clarity increases there is significant decrease in internal fragmentation and increase in open group process and group cohesiveness. The increase in internal fragmentation results in the decrease of open group process, group cohesiveness and group homogeneity. The increase in open group process results in the increase of groups cohesiveness and group homogeneity. And the increase in group cohesiveness results in the increase of group homogeneity.

Supervisory Behaviour

The supervisor is important in determining the employee's quality of work life and work performance. The effect of leadership can be studied from different perspectives. In the present study the supervisor's style for the purpose of describe him/her (supervisor) as an aspect of respondent's environment was studied. The supervisory behaviour as perceived or experienced by the subordinates was studied.

Table 8.11 shows the mean scores, standard deviation and frequency distribution of respondents on the perception of their supervisor's behaviour. It is clear from the results that out of ten dimensions of supervisor's style or functioning, the employees of the department of agriculture of the developed state differed significantly from the employees of the less developed state with respect to seven dimensions. The employees of the developed state had significantly higher score on the perception of their supervisor's style with regard to work facilitation-subordinate relations, work facilitation-goal settling, decision decentralisation, production orientation, control of work, competence and participation. The employees of both the states did not differ with respect to their perception of work facilitation-problem solving, consideration and bias.

The examination of Table 8.11 also reveals that a majority of the employees of both the states had medium score with regard to various dimensions of supervisor behaviour. However, a higher percentage of the developed state employees fell under high score than that of their counterparts in the less developed state on the perception of supervisor's style with regard to all the dimensions, except bias. The percentage of the developed state employees who fell under high score for the nine dimensions (except bias) ranged from 26.67 to 45.33 (average 36.00%) whereas for the employees of the less developed state the above range was only from 14.67 to 26.00 (average 18.44%). In contrast the percentage of the employees of the developed state who fell under low score for the above nine dimensions (except bias) ranged only from 0.00 to 28.67 (average 8.74%), whereas in the case of employees of the less developed state the range was between 14.67 and 38.67 (average 24.88%).

The above results clearly indicate that the employees of the developed state and the less developed state significantly differed from each other with respect to their perception about their supervisor's style of functioning. In general the employees of the developed state when compared to their counterparts in the less developed state perceived their supervisor or leadership effective in their functioning.

It is interesting to see that the employees of both the states did not have significant difference in their perception of their supervisors with regard to consideration, work facilitation-problem solving and bias. This means that both the groups of employees had and work

TABLE 8.10: Inter-correlations among work group dimensions

Dimensions	DS					LDS				
	1	2	3	4	5	1	2	3	4	5
1. Group goal clarity	1.00					1.00				
2. Internal fragmentation	–0.18*	1.00				–0.21**	1.00			
3. Open group process	0.29**	–0.20*	1.00			0.31**	–0.55**	1.00		
4. Group cohesiveness	0.41**	–0.61**	0.19*	1.00		0.20*	–0.68**	0.59**	1.00	
5. Group homogeneity	0.12	–0.24**	0.26**	0.36	1.00	–0.01	–0.47**	0.39**	0.71**	1.00

DS= Developed State LDS = Less Developed State

** Significant at 0.01 level of probability * Significant at 0.05 level of probability.

TABLE 8.11: Mean scores, standard deviation and frequency distribution of respondents on supervisory behaviour dimensions

Supervisory Behaviour Dimensions		Range	Mean	S.D.	High score		Medium score		Low score		Value of 't'
					No.	%	No.	%	No.	%	
1. Work facilitation-subordinate relations	DS	5-35	23.38	5.64	51	34.00	90	60.00	9	6.00	
	LDS	5-35	19.97	5.27	28	18.67	84	56.00	138	25.33	5.39**
2. Work facilitation-problem solving	DS	2-14	9.78	2.23	43	28.67	83	55.33	24	16.00	
	LDS	2-14	9.64	5.18	35	23.33	78	52.00	37	24.67	0.30NS
3. Work facilitation-goal setting	DS	2-14	10.47	2.28	63	42.00	83	55.33	4	2.67	
	LDS	2-14	9.74	1.95	33	22.00	89	59.33	28	18.67	2.95**
4. Consideration	DS	2-14	10.08	2.46	49	32.67	86	57.33	15	10.00	
	LDS	2-14	9.89	2.78	39	26.00	89	59.33	22	14.67	0.62NS
5. Decision decentralisation	DS	2-14	8.69	2.84	40	26.67	67	44.67	43	28.67	
	LDS	2-14	8.10	1.62	23	15.33	69	46.00	58	38.67	2.25*
6. Bias	DS	2-14	5.69	4.52	35	23.33	92	61.33	23	15.33	
	LDS	2-14	6.38	4.89	47	31.33	67	44.66	36	24.00	−1.27NS
7. Production orientation	DS	3-21	15.93	3.45	63	42.00	87	58.00	0	0.00	
	LDS	3-21	13.19	3.57	23	15.33	89	59.33	38	25.33	6.74**
8. Control of work	DS	4-28	19.37	4.81	64	42.67	82	54.67	4	2.67	
	LDS	4-28	15.01	5.11	22	14.67	96	64.00	32	21.33	7.58**
9. Competence	DS	1-7	4.99	1.27	46	30.67	92	61.33	12	8.00	
	LDS	1-7	4.48	1.73	18	12.00	88	58.67	44	29.33	2.87**
10. Participation	DS	2-14	9.78	2.48	68	45.33	75	50.00	7	4.67	
	LDS	2-14	8.13	2.56	28	18.67	83	55.33	39	26.00	5.67**

DS= Developed State
** Significant at 0.01 level of probability
NS= Not Significant.

LDS = Less Developed State
* Significant at 0.05 level of probability

environment which did not differ with respect to their supervisor's consideration, work facilitation-problem solving and bias. But the supervisors of the less developed state were regarded relatively less in the attributes like decision decentralisation, production orientation, control of work, competence, work facilitation through goal setting and participation. This calls for effective training to the supervisors of the less developed state on the modern principles of supervision and management. Even the field observation confirmed the above points as it was observed that many officials of the department of agriculture of the less developed state viewed their role largely as that of inspecting officers. Further it was also observed that due to the conditions of their poor official status and other organisational problems they lacked motivation to effectively guide and supervise the subordinates.

From the above results it can be concluded that the department of agriculture of the developed state differed from the less developed state with regard to the functioning of supervisors. The employees of the developed state had relatively more support in terms of effective supervision or leadership. These findings support the General Hypothesis, GH-10, which states that there will be differences in the supervisory behaviour of the officials of the department of agriculture of the developed state and the less developed state.

Inter-Correlations Among the Supervisory Behaviour Dimensions

Table 8.12 shows the inter-correlations among the supervisory behaviour dimensions. The examination of the data reveals a considerable amount of relationship among the supervisory behaviour dimensions.

In the case of both the groups of employees it is observed that a significant increase in work facilitation through subordinate relations results in significant increase in the employee's perception of supervisor's behaviour in the areas like consideration, production orientation, control of work, results also indicate that consideration was significantly and positively associated to production orientation, competence, participation, goal setting and problem solving. The decision decentralisation was positively and significantly associated to control of work and problem solving.

The correlations further indicate that significant increase in production orientation results in the increase of control of work, competence, participation and goal setting. The control of work was positively and significantly associated with competence, participation

TABLE 8.12: Inter-correlations among supervisory behaviour dimensions

Dimensions		1	2	3	4	5	6	7	8	9	10
1. Work facilitation subordinate relations	DS	1.00									
	LDS	1.00									
2. Consideration	DS	0.62**	1.00								
	LDS	0.45**	1.00								
3. Decision decentralisation	DS	0.11	0.07	1.00							
	LDS	0.02	—0.02	1.00							
4. Bias	DS	—0.10	0.04	0.08	1.00						
	LDS	—0.08	—0.09	0.03	1.00						
5. Production orientation	DS	0.53**	0.28**	—0.01	—0.17	1.00					
	LDS	0.25**	0.16*	0.13	—0.09	1.00					
6. Control of work	DS	0.27**	0.04	0.18*	0.06	0.25**	1.00				
	LDS	0.51**	—0.07	0.23**	—0.03	0.20*	1.00				
7. Competence	DS	0.29**	0.16*	—0.09	—0.16*	0.25**	0.37**	1.00			
	LDS	0.36**	0.19*	0.06	—0.13	0.34**	0.49**	1.00			
8. Participation	DS	0.50**	0.41**	0.01	—0.08	0.29**	0.47**	0.34**	1.00		
	LDS	0.34**	0.35**	0.32**	—0.16*	0.21**	0.24**	0.10	1.00		
9. Goal setting	DS	0.16*	0.18*	0.28**	—0.04	0.23**	0.24**	0.04	0.24**	1.00	
	LDS	0.02	0.29**	—0.03	—0.03	0.25**	0.20*	0.05	0.05	1.00	
10. Work facilitation-problem solving	DS	0.20*	0.19*	0.37**	—0.10	0.04	—0.07	0.01	0.11	0.02	1.00
	LDS	0.19*	0.29**	0.17*	—0.14	0.03	—0.03	—0.06	0.07	0.03	1.00

DS= Developed State LDS = Less Developed State

** Significant at 0.01 level of probability * Significant at 0.05 level of probability.

and goal setting. And the participation was positively and significantly related with goal setting.

Apart from the above relationship common to the employees of the two states, there are significant relationships found specific to the employees of the developed state, as well to the employees of the less developed state.

It can be concluded from the above results that there is a high degree of inter-correlations among the supervisory behaviour dimensions as perceived by the employees of both states. The perception of supervisory behaviour in one dimension is related to many other dimensions. This shows the high degree of relationship between supervisory dimensions studied.

Time-Use Characteristics

Table 8.13 presents the data on time-use characteristics of the employees of the department of agriculture of the developed state and the less developed state. The time use dimensions investigated in the present investigation were employee's routine, time structure, direction and engagement. It can be seen from the data that both the groups of employees differed significantly from each other in all the dimensions of time use. The employees of the developed state had significantly higher score on routine, time structure, direction and engagement than that of the employees of the less developed state.

The examination of the data also reveals that a majority of the employees of both the states had medium score on time use dimensions. But both the groups differed with respect to their distribution on high and low scores. The percentage of developed state employees who received high score on routine, time structure, direction and engagement was 36.67, 46.67, 40.67 and 30.00 respectively as against a lesser percentage of less developed employees (23.33, 19.33, 18.67 and 20.00 per cent, respectively) who fell in this category. On the contrary a relatively higher percentage of employees of the less developed state had low score (26 to 34.67%, average 30.67%) on the above dimensions when compared to the employees of the developed state (4.67 to 8.00%, average 6.17%).

It is clear from the findings that the employees of the state department of agriculture in the developed state had better time use in terms of their relatively higher routine, time structure, direction and engagement than of the employees of the less developed state. The

TABLE 8.13: Mean scores, standard deviation and frequency distribution of respondents on time use dimensions

Time use dimensions		Range	Mean	S.D.	High score		Medium score		Low score		Value of 't'
					No.	%	No.	%	No.	%	
1. Routine	DS	3-15	11.79	2.54	55	36.67	83	55.33	12	8.00	
	LDS	3-15	11.10	2.44	35	23.33	76	50.67	39	26.00	2.38*
2. Structure	DS	2-10	7.49	1.64	70	46.67	73	48.67	7	4.67	
	LDS	2-10	5.73	2.02	29	19.33	73	48.67	48	32.00	8.26**
3. Direction	DS	3-15	10.66	2.32	61	40.67	79	52.67	10	6.67	
	LDS	3-15	9.22	2.34	28	18.67	77	51.33	45	30.00	5.33**
4. Engagement	DS	4-20	14.63	2.72	45	30.00	97	64.67	8	5.33	
	LDS	4-20	12.66	3.40	30	20.00	68	45.33	52	34.67	5.51**

DS= Developed State
LDS = Less Developed State
** Significant at 0.01 level of probability
* Significant at 0.05 level of probability.

effectiveness in use of time reflects the effectiveness in managing the work and the management of self itself. The employees of the less developed state had showed relatively less routine, structure, direction and engagement in the use of time. In general, their use of time indicate less organisation of the work and less purposeful activities. This may be the result of relatively poor job-design and leadership. As expressed by some respondents of the less developed state, the trouble in planning and organising the work was mainly due to lack of effective planning, monitoring of work and the tendency to leave the things until the last minute. It was also revealed in the field observations that though these employees had a rough plan about the activities to be done, it was upset by the intervention from the supervisors and unscheduled activities. Further the problems with the organisation of time might be due to the ineffective supervision and guidance. In contrast it was observed that in the developed state effective periodical review of the work done through the fortnightly training sessions or monthly review meetings the employees in having more direction and organisation of work.

TABLE 8.14: Inter-correlations among time use dimensions

Time use dimensions	DS				LDS			
	1	2	3	4	1	2	3	4
1. Routine	1.00				1.00			
2. Structure	0.17*	1.00			0.20**	1.00		
3. Direction	0.16*	0.18*	1.00		0.17*	0.18*	1.00	
4. Engagement	0.16*	0.23*	0.21**	1.00	0.14*	0.16*	0.25**	1.00

DS= Developed State
LDS = Less Developed State
** Significant at 0.01 level of probability
* Significant at 0.05 level of probability

From the above results, it can be concluded that the employees of the department of agriculture of the developed state and the less developed state significantly differed from each other with regard to time use characteristics. The employees of the developed state were relatively better in time use in terms of higher routine, time structure, direction and engagement. This support the General Hypothesis, GH-11, which states that there will be differences in the time use characteristics of the employees of the department of agriculture of the developed state and the less developed state.

Inter-Correlations Among the Time-Use Dimensions

Table 8.14 shows the inter-correlations among the time use dimensions. The results of correlations for the both groups of employees show that the routine (routine and planning) is positively and significantly associated with structure, direction and engagement. As expected, the increase in routine results in the significant increase in the structure (organisation of work) as well as direction (direction, having aim and easiness in initiating work) and engagement (having purpose, meaning and interest). Likewise the structure is positively and significantly associated with direction and engagement. The increase in structuring of time (organisation) results in the significant increase in the direction and engagement and the direction was positively and significantly associated with engagement. This means that the increase in direction results in significant increase in engagement.

Job Technology

Table 8.15 presents the mean scores, standard deviation and frequency distribution of respondents in job technology scores. The following aspects of job technology were studied in the present investigation : (1) job dependence among unit personnel, (2) predictability of events in the job, (3) routineness of operation, and (4) job inputs adequacy

The data in Table 8.15 reveals that the employees of department of agriculture in the developed state significantly differed from the employees of the less developed state with regard to predictability of events in the job, routineness of operation, and job inputs adequacy. The employees of the developed state had significantly higher score than that of employees of the less developed state on routineness of operation and job inputs adequacy but lesser score on predictability of events.

As the data further indicate that most of the employees of both groups had medium score on job technology dimensions. But a considerable percentage of employees of the developed state had high score on dimensions like routineness of operation (33.66%) and job inputs adequacy (37.33%) which was higher than that of percentage of employees of less developed state who fell under this category (18% and 8% respectively). In contrast a higher percentage of employees of the less developed state fell under low score on job inputs adequacy (31.33%) than their counterparts in the developed state (4.66%).

TABLE 8.15: Mean scores, standard deviation and frequency distribution of respondents on job-technology scores

Dimensions		Range	Mean	S.D.	High score		Medium score		Low score		Value of 't'
					No.	%	No.	%	No.	%	
1. Job dependence	DS	3-15	8.29	2.16	37	24.67	84	56.00	29	19.33	
(on supervisors)	LDS	3-15	8.82	2.56	49	32.67	82	54.67	19	12.67	–0.74 NS
2. Job dependence	DS	3-15	8.81	2.64	15	10.00	96	64.00	39	26.00	
(on members)	LDS	3-15	8.47	3.04	5	3.33	97	64.66	48	32.00	1.03 NS
3. Job dependence	DS	3-15	5.43	2.97	27	18.00	71	47.33	52	34.67	
(on others)	LDS	3-15	4.92	2.67	20	13.33	63	42.00	67	44.67	1.56 NS
4. Predictability of	DS	2-10	6.19	1.59	33	22.00	89	59.33	28	18.67	
events	LDS	2-10	6.79	1.53	49	32.67	89	59.33	12	8.00	–3.28**
5. Routineness of	DS	2-10	7.29	1.46	49	32.66	70	46.66	31	20.67	
operation	LDS	2-10	6.63	1.66	27	18.00	108	72.00	15	10.00	3.61**
6. Job inputs adequacy	DS	3-15	9.68	2.07	56	37.33	87	58.00	7	4.66	
	LDS	3-15	8.07	2.18	12	8.00	91	60.66	47	31.33	6.54**

DS= Developed State
LDS = Less Developed State
** Significant at 0.01 level of probability
* Significant at 0.05 level of probability
NS= Not Significant.

A considerable percentage of employees of both the groups had low score on job dependence among unit members (DS = 26.00%, LDS = 32.00%) and job dependence on external unit members (DS = 34.67%, LDS = 44.67%).

The employees of both the states did not significantly differ on job dependence among unit personnel (supervisor, members and other unit members). It means that the work interconnectedness and the intensity of work interdependence did not vary between the employees of the department of agriculture of the developed state and the less developed state.

The employees of the less developed state had a significantly higher score on predictability of events than that of their counterparts in the developed state. This means that the employees of the less developed state were able to anticipate and predict the nature of events pertaining to their job in a greater measure than that of employees of the developed state. They perhaps encountered the same kind of problems to be solved in work situation. But on the other hand the employees of developed state had to face more new situations and new problems to be tackled. This might be due to the complexity of developed agriculture in the developed state. Though the employees of the developed state had relatively low score in predictability of events than that of less developed state, they had significantly higher score in routineness of operation. Their procedure for field work or office administration was relatively more standardised with a regular routine of work to be followed. The higher score in this area might be due to their higher unit and job standardisation. As expressed by many officials, the lower routineness of operation of employees of the less developed state was due to lack of standardised operational and administrative procedures.

The employees of the less developed state had significantly low score on job inputs adequacy than that of the employees of the developed state. This means that the resources for performance of job in terms of availability of materials, facilities and information, were relatively lacking more in the department of agriculture of the less developed state. One of the serious problems with regard to job inputs adequacy in the less developed state as evidenced by field observation was lack of transport at field level. The employees also complained of inadequate fixed travelling allowance. Further as expressed by many officials, the continuous flow of information on latest technology, programmes, availability of literature and other farm inputs which would help them in the performance of job was hindered due to the problems in co-ordination, ineffective training sessions, inadequate finance and non-availability of subject matter specialists. The farm inputs needed by the field functionaries in the developed state were easily available to them since a considerable amount of inputs especially seeds and pesticides, was distributed through their departmental supply depots which were located at Block level. In contrast the supply of above inputs in less developed state was mainly done by co-operatives or other agencies and this arrangement had caused problems in co-ordination resulting in inadequate supply or untimely supply of inputs.

From the above results and discussion it can be concluded that the job technology of the department of agriculture of the developed state significantly differed from that of the less developed state. The employees of the agricultural department of the developed state were characterised by having significantly lesser predictability of events related to their work, but a higher routineness of operation and job inputs adequacy. This supports the General Hypothesis, GH-12, which states that there will be differences in the job technology of the department of agriculture of the developed state and the less developed state.

Inter-Correlations Among the Job-Technology Variables

Table 8.16 shows the correlations among the job technology variables. The results of correlations show that a decrease in job dependence on members as well as job inputs adequacy results in increased job dependence on supervisors. An increased job dependence on supervisors as well as on others (outside the department) results in decreased predictability of events related to the job. Further the correlational analysis reveals that the decreased job inputs adequacy results in increased job dependence on the members as well as supervisors of the work group. Predictability of events was positively and significantly associated with routineness of operation. Thus it is clear a higher routine jobs results in the higher predictability of events related to the work.

Pay Administration

Table 8.17 shows the mean scores, standard deviation and frequency distribution of the respondents on the pay administration dimensions. It can be seen from the data that the pay administration of the employees of the developed state significantly differed from the employees of the

TABLE 8.16: Inter-correlations among job-technology variables

Variables	DS						LDS					
	1	2	3	4	5	6	1	2	3	4	5	6
1. Job dependence (on supervisors)	1.00						1.00					
2. Job dependence (on members)	–0.16*	1.00					–0.18*	1.00				
3. Job dependence (on others)	0.08	0.05	1.00				–0.09	0.16*	1.00			
4. Predictability of events	–0.18*	–0.21**	–0.17*	1.00			–0.22**	–0.05	–0.19*	1.00		
5. Routineness of operation	0.12	0.08	–0.01	0.49**	1.00		–0.14	0.01	0.02	0.54**	1.00	
6. Job inputs adequacy	–0.16*	–0.23**	0.02	0.01	–0.12	1.00	–0.21*	–0.19*	0.10	0.04	0.11	1.00

DS= Developed State
** Significant at 0.01 level of probability

LDS = Less Developed State
* Significant at 0.05 level of probability

TABLE 8.17: Mean scores, standard deviation and frequency distribution of respondents on pay administration dimensions

Pay administration dimensions		Range	Mean	S.D.	High score		Medium score		Low score		Value of 't'
					No.	%	No.	%	No.	%	
1. Pay administration satisfaction	DS	2-14	8.52	3.26	66	44.00	69	46.00	15	10.00	5.99**
	LDS	2-14	6.63	2.07	48	32.00	67	44.67	35	23.33	
2. Pay satisfaction	DS	3-21	8.95	2.73	28	18.67	86	57.33	36	24.00	2.54**
	LDS	3-21	8.17	2.57	17	11.33	76	50.66	57	38.00	
3. Personal equity	DS	3-21	8.61	2.83	41	27.33	80	53.33	29	19.33	2.75**
	LDS	3-21	7.70	2.80	26	17.33	76	50.67	48	32.00	
4. External equity	DS	2-14	7.61	2.94	38	25.33	81	54.00	31	20.67	4.52**
	LDS	2-14	6.17	2.55	18	12.00	70	46.67	62	41.33	
5. Performance-pay contingency	DS	2-14	7.82	2.52	43	28.67	89	59.33	18	12.00	3.21**
	LDS	2-14	6.87	2.60	19	12.67	86	57.33	45	30.00	

DS= Developed State

** Significantly at 0.01 level of probability

LDS = Less Developed State

* Significant at 0.05 level of probability.

less developed state with regard to all dimensions. The employees of the developed state had significantly higher score on pay administration satisfaction, pay satisfaction, personal equity, external equity, and performance pay contingency.

The data further reveals that a majority of the employees of both the states had medium score on the pay administration dimensions. A relatively higher percentage of the developed state employees had high score on dimensions like pay administration satisfaction, pay satisfaction, personal equity, external equity and performance pay contingency. The percentage of the developed state employees who fell under high score category of the above five dimensions ranged from 18.67 to 44.00 (average 28.80%) whereas in the case of employees of the less developed state the above range was only from 11.33 to 32.00 (average 17.06%). In contrast the percentage of the developed state employees who fell under the low score for the above five dimensions ranged from 10.00 to 24.00 (average 17.2%) whereas in the case of employees of the less developed state the above range was from 30.00 to 41.33 per cent (average 32.93%).

The examination of the results indicate that the over all pay administration as perceived by the employees was relatively better in the case of department of agriculture of the developed state. As indicated earlier employees of the developed state had significantly higher score on all aspects of pay administration dimensions. A significantly higher score of the developed state employees on pay administration satisfaction indicates that the aspects of pay administration like timely payment of salary, annual increment, dearness allowance and the like are relatively better in the department of agriculture of the developed state than that of its counterpart in the less developed state.

The relatively low pay satisfaction of the employees of the less developed state might be due to the fact that the initial salary of the employees of the less developed state especially of agricultural extension officers were lower than that of their counterparts in the developed state. Further the employees of the developed state had the advantage of having selection grades which would also increase their pay satisfaction. The employees of the less developed state had significantly lower score on external equity and personal equity with regard to their pay than that of the employees not equitable in comparison to the pay of their counterparts in other departments or to their own inputs on the job. This was due to the fact that the pay scale of employees of the department of agriculture in the less developed state was lower when compared to the pay scale of their counterparts in other departments like Animal Husbandry, Public Works and Health. In contrast in the developed state the pay scale of the employees of department of agriculture was equivalent to the pay scale of their counterparts in other departments like Animal Husbandry and Public Works. For example, the Agricultural Officer (equivalent to A.E.O. in the Blocks) received the pay equivalent to that of Junior Engineer or Assistant Veterinary Surgeon in the developed state. In contrast the pay scale of Agricultural Development Officer (equivalent to A.E.O. in the Block) was much lower than that of Junior Engineer or Assistant Veterinary Surgeon in the less developed state.

It was interesting to see that the performance pay contingency score of the employees of the developed state was higher than that of the employees of the less developed state. This means that the employees of the developed state had a greater degree of perception that the pay contingent on their level of performance on the job, though in general the pay is not contingent upon the performance in government departments. The pay increase and the promotion generally depend upon seniority.

From the above results it can be concluded that the dimensions of pay administration as perceived by the employees of the department of agriculture of the developed state and the less developed state differed from each other. The employees of the department of agriculture of the developed state had significantly higher level of pay satisfaction, personal equity, external equity and performance pay contingency. The findings support the General Hypothesis, GH-13, which states that there will be differences in pay administration dimensions between the department of agriculture of the developed state and the department of agriculture of the less developed state.

Inter-Correlations Among Pay Administration Dimensions

Table 8.18 shows the inter-correlations among the pay administration variables. It shows a considerable amount of inter-correlations among the pay administration variables.

In the case of both the groups of employees the correlations indicate that the pay administration satisfaction was positively and significantly associated with pay satisfaction, personal equity and external equity. Thus an increase in pay administration satisfaction results in the significant increase in pay satisfaction, personal equity and external equity.

TABLE 8.18: Inter-correlations among pay administration dimensions

Dimensions	DS					LDS				
	1	2	3	4	5	1	2	3	4	5
1. Pay administration satisfaction	1.00					1.00				
2. Pay satisfaction	0.35**	1.00				0.51**	1.00			
3. Personal equity	0.44**	0.69**	1.00			0.23**	0.43**	1.00		
4. External equity	0.38**	0.21**	0.14	1.00		0.44**	0.64**	0.18*	1.00	
5. Performance pay contingent	–0.09	–0.14	0.07	–0.09	1.00	–0.13	–0.03	0.07	–0.08	1.00

DS= Developed State LDS = Less Developed State

** Significant at 0.01 level of probability * Significant at 0.05 level of probability.

The pay satisfaction was positively and significantly associated with personal equity and external equity. This means that those employees who had higher pay satisfaction also had higher personal and external equity. The increase in pay satisfaction results in the increase of external equity and personal equity.

The results of correlations also show that among the employees of the less developed state (not found among the employees of the developed state) the personal equity was positively and significantly associated with external unity. This means that among the employees of the less developed state those who perceived a higher personal equity also perceived a higher external equity.

The overall picture of the correlations shows that there is a considerable amount of inter-correlations among the dimensions of pay administration.

Creativity Climate

Creativity is the ability to think fast, generate new ideas and deal with problems in an effective way. Creative abilities are needed by all people irrespective of the types of job they perform, though some jobs need more creative abilities than other jobs. The application of creative thinking in the organisation will make it more productive and efficient in solving problems. The creativity is a complex process and its outcome depends upon the person's personality traits, abilities, as well as his environment. Various research findings have clearly pointed out the importance of environment in stimulating creativity. In the present investigation the creativity environment of the department of agriculture in the developed state as well as the less developed state was studied.

Table 8.19 shows the mean scores, standard deviation and frequency distribution of the respondents on their perception of creativity environments. The examination of the data presented in Table 8.19 reveals that the perceived creativity climate of the department of agriculture in the developed state differed from that of the less developed state with regard to five out of eight dimensions of creativity. The employees of the developed state perceived significantly higher score on availability of constructive feed back, degree of stimulation, encouragement and reward to creativity, opportunities for higher training and availability of physical facilities for experimentation. Both the groups of employees did not differ significantly with regard to the dimensions like absence of premature criticism, freedom of thought and availability of diverse view points.

A persual of the data further reveals that a majority of the respondents had medium score on the creativity climate. But a higher percentage of the employees of the developed state had high score in all the dimensions of creativity climate than that of the employees of the less developed state. The percentage of the developed state employees who had high score for all the eight dimensions ranged from 14.67 to 43.33 with an average of 26.83. In the case of the employees of less developed state the above range was from 7.33 to 25.33 per cent with an average of 14.42. The percentage of the developed state employees who had low scores on the eight dimensions of creativity climate ranged from 6.67 to 24.67 with an average of 16.65. A relatively higher percentage of the less developed state employees had low scores in the above dimensions of creativity climate, which ranged from 21.33 to 43.33 with an average of 29.58.

The above results clearly indicate that the creativity environment of the department of agriculture of the developed state significantly differed from that of the less developed state. The employees of the less developed state had relatively less score on creativity climate dimensions like availability of constructive feed back, degree of stimulation, encouragement and reward to creativity, opportunities for higher training, and availability of physical facilities for experimentation. The reasons for this state of affair might be due to their conformity to the rules and regulations and general culture of the organisation which is satisfied with status quo and discourages any new ideas or practices related to the work. In general, the creativity climate in government department is poor and culture of any government organisation mostly discourages creative ideas and encourages only to follow precedence. Under such circumstances the creative individuals will only be discouraged. The relatively high score of employees of the developed state on creativity climate indicates that the department of agriculture of developed state has undergone considerable changes in its functioning when compared to its counterparts in the less developed state, which has resulted in creating an environment which is favourable for creative ideas.

It can be concluded from the results of the study that the creativity climate of the department of agriculture of the developed state significantly differed from that of the less developed state. The overall creativity climate as perceived by the employees of the developed state was relatively better than that of the employees of the less developed state. The findings support the General Hypothesis, GH-14, which

TABLE 8.19: Mean scores, standard deviation and frequency distribution of respondents on creativity climate

Creativity climate dimensions		Range	Mean	S.D.	High score No.	High score %	Medium score No.	Medium score %	Low score No.	Low score %	Value of 't'
1. Absence of premature criticism	DS	1-5	2.94	1.10	45	30.00	80	53.33	25	16.67	
	LDS	1-5	2.70	1.50	38	25.33	74	49.33	38	25.33	1.58 NS
2. Availability of constructive feed back	DS	1-5	2.91	1.07	53	35.33	83	55.33	14	9.33	
	LDS	1-5	2.53	1.06	28	18.67	85	56.67	37	24.67	3.04**
3. Degree of stimulation	DS	1-5	3.15	1.07	65	43.33	75	50.00	10	6.67	
	LDS	1-5	2.53	0.97	19	12.67	89	59.33	42	28.00	5.21**
4. Freedom of thought	DS	1-5	2.53	1.25	32	21.33	86	57.33	32	21.33	
	LDS	1-5	2.24	1.60	25	16.67	86	57.33	39	26.00	1.75 NS
5. Encouragement and reward of creativity	DS	1-5	2.65	1.05	38	25.33	91	60.67	21	14.00	
	LDS	1-5	2.18	0.88	12	8.00	94	62.67	44	29.33	4.20**
6. Opportunities for higher training	DS	1-5	2.26	0.99	28	18.67	85	56.67	37	24.67	
	LDS	1-5	1.89	0.85	11	7.33	81	54.00	58	38.67	3.43**
7. Availability of physical facilities for experimentation	DS	1-5	1.96	0.99	39	26.00	88	58.67	23	15.33	
	LDS	1-5	1.73	0.74	22	14.67	63	42.00	65	43.33	2.36*
8. Availability of diverse view points	DS	1-5	2.25	1.02	22	14.67	91	60.67	37	24.67	
	LDS	1-5	2.09	0.85	18	12.00	100	66.67	32	21.33	1.48 NS

DS= Developed State
** Significant at 0.01 level of probability
NS= Not Significant.

LDS = Less Developed State
* Significant at 0.05 level of probability

states that there will be differences between the creativity climate of the department of agriculture of the developed state and the less developed state.

Inter-Correlations Among the Dimensions of Creativity Climate

Table 8.20 shows the inter-correlations among the creativity climate dimensions. The results of correlations show a fairly high amount of positive inter-correlations of moderate size among the creativity climate dimensions.

With regard to both the groups of employees the correlations indicate that absence of premature criticism was positively and significantly associated with availability of feed back, freedom of thought and action, encouragement and reward to creativity and availability of diverse view points. This means an increase in the absence of premature criticism to creative ideas results in higher feed back, freedom of thought and action, encouragement and reward to creativity and availability of diverse view points.

Availability of constructive feed back on the creative efforts was positively and significantly correlated with freedom of thought and availability of diverse view points. Thus, it is clear that an increased constructive feed back for the creative efforts results in significant increase in the freedom of belief and availability of diverse view points.

Degree of stimulating environment was positively and significantly associated with encouragement and reward to creativity efforts. Thus an increase in the stimulating environment results in the significant increase in the encouragement and reward to creativity.

Likewise freedom of belief and action was positively and significantly correlated with availability of diverse view points. And encouragement and reward to creativity was positively and significantly associated with opportunities for higher learning.

Correlations in Table 8.19 also show that among the employees of the developed state (not found among the employees of the less developed state) the availability of constructive feed back was positively and significantly correlated with degree of stimulating environment for creativity and opportunities for higher learning. This means that among the employees of the developed state an increase in the constructive feed back results in the increase of stimulating environment for creativity as well opportunities for higher learning.

TABLE 8.19: Inter-correlations among creativity climate dimensions

Dimensions		1	2	3	4	5	6	7	8
1. Absence of criticism	DS	1.00							
	LDS	1.00							
2. Availability of feed back	DS	0.49**	1.00						
	LDS	0.44**	1.00						
3. Degree of stimulation	DS	0.03	0.18*	1.00					
	LDS	0.09	–0.02	1.00					
4. Freedom of thought	DS	0.38**	0.27**	0.05	1.00				
	LDS	0.41**	0.18*	–0.03	1.00				
5. Encouragement and reward to creativity	DS	0.24**	–0.08	0.22**	0.23**	1.00			
	LDS	0.25**	0.11	0.31**	–0.01	1.00			
6. Opportunities for higher learning	DS	0.08	0.21**	0.06	0.10	0.25**	1.00		
	LDS	–0.02	–0.01	0.07	–0.14	0.28**	1.00		
7. Availability of physical facilities	DS	–0.01	0.11	–0.04	0.01	0.07	0.14	1.00	
	LDS	0.00	0.12	0.08	0.02	–0.05	0.01	1.00	
8. Availability of diverse view points	DS	0.38**	0.32**	0.29**	0.50**	0.07	0.26**	0.10	1.00
	LDS	0.27**	0.27**	0.07	0.38**	–0.01	0.10	0.05	1.00

DS= Developed State LDS = Less Developed State

** Significant at 0.01 level of probability

* Significant at 0.01 level of probability.

Likewise the degree of stimulation was significantly and positively associated with availability of diverse view points. An increase in the stimulating environment for creativity also results in the increased availability of diverse view points.

The freedom of belief and action was positively and significantly associated with the availability of encouragement and reward to creativity. This means that among the employees of the developed state an increase in higher freedom of belief and action results in a higher availability of encouragement and reward to creativity.

The opportunities of higher learning was positively and significantly associated with availability of diverse view points. This means among the employees of the developed state an increase in the opportunities for higher learning also results in increased availability of diverse view points.

It can be concluded from the above results that there is a high degree of relationship among the dimensions of creativity climate. The employees perception of one dimension is closely related to the perception of other dimensions. Thus an increase in the creativity climate of one dimension results in the increase of other dimensions.

9

ORGANISATIONAL EFFECTIVENESS

The organisational effectiveness of the department of agriculture was measured in the study with respect to the following dimensions : (1) Perception of organisational performance, (2) Organisational adaptability-flexibility, (3) Job satisfaction of the employees, and (4) Organisational involvement of the employees.

Perception of Organisational Performance

The perception of the employees of the department of agriculture about the performance of their department was considered as one of the measures of effectiveness of the department of agriculture in the study. Table 9.1 presents the mean scores, standard deviation and frequency distribution of respondents on various dimensions of performance. The examination of the data the Table 9.1 shows that the overall mean performance score of the department of agriculture of developed state (42.88) was significantly higher than that of the department of agriculture in the less developed state (37.53).

The department of agriculture of the developed state had the highest performance score on the supply of inputs (5.32) followed by training of the employees (5.23), provision of advisory services (4.58), training of farmers (4.53), achievement of target (4.38), reputation for work excellence (4.38), quality of work done (4.21), regularity in field visit (3.95), total quantity of work done (3.25) and conducting adaptive trials and minikit programme (3.05). The department of agriculture of the less developed state received the highest mean performance score on achievement of target (4.68) followed by training of employees (4.61), training of farmers (4.08), provision of advisory services (3.83), quality of work done (3.63), supply of inputs (3.60), reputation for work excellence (3.45), regularity in field visit (3.37), conducting adaptive trials and minikit programmes (3.21) and total quantity of work done (3.07).

TABLE 9.1: Mean scores, standard deviation and frequency distribution of respondents on their perception of performance of department of agriculture

Performance dimensions		Range	Mean	S.D.	High score		Medium score		Low score		Value of 't'
					No.	%	No.	%	No.	%	
1	2	3	4	5	6	7	8	9	10	11	12
1. Reputation for	DS	1-7	4.38	1.21	53	35.33	73	48.67	24	16.00	
work excellence	LDS	1-7	3.45	1.37	12	8.00	82	54.67	56	37.33	6.23**
2. Quality of work	DS	1-7	4.21	1.23	51	34.00	80	53.33	19	12.67	
done	LDS	1-7	3.63	1.19	16	10.67	85	56.67	49	32.67	4.15**
3. Total quantity of	DS	1-7	3.25	1.31	28	18.66	74	49.33	48	32.00	
work done	LDS	1-7	3.07	1.61	18	12.00	70	46.67	62	41.33	1.06[NS]
4. Achievement of	DS	1-7	4.38	1.06	39	26.00	85	56.67	26	17.33	
target	LDS	1-7	4.68	1.61	49	32.67	89	59.33	12	8.00	–1.91[NS]
5. Provision of	DS	1-7	4.58	1.03	48	32.00	88	58.67	14	9.33	
advisory services	LDS	1-7	3.83	1.23	17	11.33	91	60.66	42	28.00	5.83**
6. Supply of inputs	DS	1-7	5.32	1.28	68	45.33	76	50.67	6	4.00	
	LDS	1-7	3.60	1.42	20	13.33	79	52.66	51	34.00	11.09**

(Contd.)

TABLE 9.1 *(Contd.)*

1	2	3	4	5	6	7	8	9	10	11	12
7. Training of farmers	DS	1-7	4.53	1.12	62	41.33	79	52.66	9	6.00	
	LDS	1-7	4.08	1.30	18	12.00	76	50.66	56	37.33	3.21**
8. Training of employees	DS	1-7	5.23	1.09	54	36.00	85	56.67	11	7.33	
	LDS	1-7	4.61	2.21	13	8.67	90	60.00	47	31.33	3.08**
9. Conducting adaptive trials and minikit programme	DS	1-7	3.05	1.20	14	9.33	88	58.67	48	32.00	
	LDS	1-7	3.21	1.21	10	6.67	72	48.00	68	45.33	1.15^{NS}
10. Regularity in field visit	DS	1-7	3.95	0.91	49	32.66	80	53.33	21	14.00	
	LDS	1-7	3.37	1.31	12	8.00	96	64.00	42	28.00	4.45**
Overall performance score	DS	7-70	42.88	5.99	51	34.00	75	50.00	24	16.00	
	LDS	7-70	37.53	6.90	19	12.66	81	54.00	50	33.33	7.17**

DS= Developed State
** Significant at 0.01 level of probability
NS= Not Significant

LDS = Less Developed State
* Significant at 0.05 level of probability

The results indicate that in general both the departments of agriculture were rated as having relatively high performance in training of extension personnel, training of farmers, achievement of target and provision of advisory services. But in items like conducting adaptive trials and minikit programmes, quantity of work done and regularity or sincerity in field visit, both the departments received relatively lower performance score.

The department of agriculture of the developed state received the highest performance score on supply of inputs (5.32) to the farmers, while for this item the less developed state was rated as having fifth position with a score of 3.60. This was due to the fact that in the developed state a part of the inputs supply especially seeds and pesticides was done through the depots of the department of agriculture, while the arrangement for supply of inputs in the less developed state was mainly through co-operatives. Supply of inputs through co-operatives and other agencies faced the problems of co-ordination, untimely supply and inadequate supply. Achievement of target seems to be the main concern of the department of agriculture of less developed state as it received the highest score (4.68) in this area.

The data in Table 9.1 indicate that out of the ten dimensions of performance, the department of agriculture of the developed state differed significantly from that of less developed state with regard to seven dimensions. The employees of the developed state had significantly higher performance score on reputation of work excellence, quality of work done, provision of advisory services, supply of inputs, training of farmers, training of the employees, and regularity in field visit. The department of agriculture of both the states did not differ with respect to their performance on total quantity of work done, achievement of target and conducting adaptive trials and minikit programmes.

The examination of the data further reveals that a majority of employees of both the states had perceived medium score on the various dimensions of performance. However, a higher percentage of the developed state employees fell under high score than that of their counterparts in the less developed state in all the dimensions of performance except 'achievement of target'. The percentage of the developed state employees who had high perception of performance score for all the dimensions ranged from 9.33 to 45.33 per cent with an average of 31.06 per cent. The above range for the employees of the less developed state was 6.67 to 32.67 with an average of 12.33 per cent. In contrast the percentage of the developed state employees who had perceived low score for all the dimensions of performance ranged from 4.00 to 32.00 with an average of 15.06 per cent whereas in the case of employees of less developed state the above range was between 8.00 and 45.33 with an average of 32.33 per cent.

From the findings of the study it can be concluded that the performance of the department of agriculture of the developed state was better than that of less developed state and it significantly differed from it.

Adaptability-Flexibility

Table 9.2 shows the findings on perception of employees of the developed and the less developed state on adaptability-flexibility of the department of agriculture. The employees of the developed state perceived that they were relatively effective in anticipating problems (4.00) followed by quickness in adopting new technology (3.27), awareness about potential new technology (3.04) and adjusting their routines to cope up with emergencies (2.45). The employees of the less developed state perceived that they are relatively effective in anticipating problems (3.41) followed by coping with emergencies (2.45), quick in adopting new technology (2.31) and awareness about potential technology (2.17). It is interesting to see that both the states were relatively effective in anticipating problems. This shows that both the departments have adequate arrangements for foreseeing or anticipating problems related to agricultural production and taking measures to solve them. Both the departments have been perceived by their employees as relatively less effective in keeping abreast of new technologies related to agricultural production or overall administration of programmes. This shows the general lack of effective linkage between agricultural departments and agricultural research institutes or universities. Further it also shows that agricultural departments being government organisations are slow to become aware about the modern techniques of management which will be useful in administration of programmes.

The data in Table 9.2 shows that the employees of the department of agriculture of the developed state significantly differed from their counterparts in the less developed state with regard to their perception of adaptability and flexibility of the department of agriculture. The department of agriculture of the developed state was rated to have higher effectiveness in all the four measures of adaptability and flexibility than that of the department of agriculture in less developed state.

TABLE 9.2: Mean score, standard deviation and frequency distribution of respondents on their perception of adaptability-flexibility of department of agriculture

Adaptability-Flexibility measures		Range	Mean	S.D.	High score No.	High score %	Medium score No.	Medium score %	Low score No.	Low score %	Value of 't'
1. Adaptability-Anticipating problems	DS	1-7	4.00	0.81	52	34.67	89	59.33	9	6.00	
	LDS	1-7	3.41	1.03	22	14.67	85	56.67	43	28.67	6.52**
2. Adaptability-awareness of potential technology	DS	1-7	3.04	1.17	62	41.33	70	46.67	18	12.00	
	LDS	1-7	2.17	0.85	12	8.00	90	60.00	48	32.00	7.36**
3. Adaptability-quickness in adopting new technology	DS	1-7	3.27	1.10	72	48.00	68	45.33	10	6.67	
	LDS	1-7	2.31	0.86	12	8.00	75	50.00	63	42.00	8.41**
4. Flexibility-coping with emergencies	DS	1-7	2.86	0.92	36	24.00	86	57.33	28	18.67	
	LDS	1-7	2.45	0.99	29	19.33	83	55.33	38	25.33	3.62**

DS= Developed State LDS = Less Developed State

** Significant at 0.01 level of probability

The data further reveal that most of the employees of both the departments had perceived medium score with regard to all the measures of adaptability and flexibility. However, a higher percentage of employees of the developed state perceived high score on various adaptability-flexibility measures (24.00 to 48.00%, average 37%) than that of employees of the less developed state (8.00 to 19.33%, average 12.5%). In contrast a higher percentage of employees of the less developed had perceived low score on various measures of adaptability-flexibility (28.67 to 42.00%, average 32%) than that of employees of the developed state (6.00 to 18.67%, average 10.83%).

From the above findings it can be concluded that department of agriculture of the developed state had higher measures of adaptability and flexibility and it significantly differed from the department of agriculture of the less developed state.

Job Satisfaction and Organisational Involvement

Job satisfaction and organisational involvement were also considered as one of the measures of effectiveness in the study. Table 9.3 presents the mean, standard deviation and frequency distribution of respondents on the above measures of effectiveness. It can be seen from the data of Table 9.3 that the mean job satisfaction score of employees of department of agriculture of the developed state was 12.75 and that of the less developed state was 11.47. Though the overall job satisfaction of employees of the developed state was little higher than that of the less developed state the difference was not statistically significant. Thus it is clear that the employees of both the states did not differ with regard to their job satisfaction.

The data also reveal that the mean score of organisational involvement of employees of the developed state was higher (10.32) than that of the less developed state (9.11). The difference was statistically significantly. Thus it is clear that the employees of the department of agriculture of the developed state were much more personally concerned and involved with the programmes of the department of agriculture than that of their counterparts in the less developed state.

The measures of effectiveness show that the department of agriculture of the developed state had significantly higher score on performance adaptability and flexibility than that of the department of agriculture in the less developed state. Further the employees of the developed state had significantly higher organisational involvement than that of their counterparts in the less developed state though both

TABLE 9.3: Mean score, standard deviation and frequency distribution of respondents on job satisfaction and organisational involvement

Performance dimensions		Range	Mean	S.D.	High score		Medium score		Low score		Value of 't'
					No.	%	No.	%	No.	%	
1. Job satisfaction	DS	3-21	12.75	5.50	18	12.00	104	69.33	28	18.67	
	LDS	3-21	11.47	7.41	9	6.00	101	67.33	40	26.67	1.69[NS]
2. Organisational involvement	DS	3-21	10.32	2.65	41	27.33	91	60.67	18	12.00	
	LDS	3-21	9.11	2.60	17	11.33	86	57.33	47	31.33	3.97**

DS= Developed State
** Significant at 0.01 level of probability

LDS = Less Developed State
NS = Not Significant

TABLE 9.4: Inter-correlations among the organisational effectiveness measures

Effectiveness Measures	DS					LDS				
	1	2	3	4	5	1	2	3	4	5
1. Job satisfaction	1.00					1.00				
2. Organisational involvement	0.36**	1.00				0.23**	1.00			
3. Adaptability (overall score)	–0.11	0.08	1.00			–0.09	0.13	1.00		
4. Flexibility (overall score)	–0.07	0.07	0.28**	1.00		0.12	0.10	0.20*	1.00	
5. Perception of organisational performance (overall score)	0.14	0.12	0.36**	0.37*	1.00	0.11	–0.08	0.27	0.18*	1.00

DS= Developed State
** Significant at 0.01 level of probability

LDS = Less Developed State
* Significant at 0.05 level of probability

the groups had similar level of job satisfaction. Thus it can be concluded that overall organisational effectiveness of the department of agriculture of the developed state was significantly higher than that of its counterpart in the less developed state. These findings support the General Hypothesis, GH-15, which states that there will be differences in the organisational effectiveness of the department of agriculture of the developed state and the less developed state.

Inter-Correlations Among the Organisational Effectiveness Measures

Table 9.4 shows the inter-correlations among the various measures of organisational effectiveness. The correlations among these measures are mostly positive and moderate in size. Thus they are not mere reproductions of one another but appear to measure different, but related organisational measures of effectiveness.

The results of correlation shows that job satisfaction was positively and significantly related to organisational involvement. It was not related to any other measures of effectiveness. Thus it is clear that an increase in the job satisfaction results in the increase of organisational involvement.

Adaptability was positively and significantly associated with flexibility and organisational performance. And flexibility was also found to be positively and significantly associated with organisational performance. Thus it is clear that the above three measures of organisational measures are related to one another the increase in one results in increase of another.

10

DISCRIMINANT ANALYSIS

A comparative account of employees of the developed as well the less developed state with regard to various socio-personal and organisational characteristics has been discussed in the foregoing sections. The results of 't' test in this regard indicated that both the groups of employees significantly differed from each other in many characteristics. The 't' values give only an idea of the statistical significance of the difference between the pairs of groups in respect of the variables but they failed to give an idea of how much they are responsible for bringing about discrimination between the comparative groups. Therefore, it was decided to subject the data for discrimination analysis to find out the percentage contribution of selected variables to the total distance measured. As both the groups of employees significantly differed from each other with regard to a number of characteristics it was felt necessary to restrict the discriminant analysis to a selected groups of variables to have a meaningful result. It was decided to study the percentage contribution of variables related to characteristics like work group functioning, supervisory functioning, time use, pay administration, creativity climate and flexibility-adaptability in discriminating the employees of state department of agriculture in the developed state from the less developed state. In all, 26 variables were included for the discriminant analysis. Discriminant analysis was done between the following groups of employees of developed as well as less developed state:

1. Combined employees of the developed state versus combined employees of the less developed state.
2. Village Level Workers (VLWs) of the developed state versus VLWs of the less developed state.
3. Junior Level Agricultural Officer (JLAO) of the developed state versus JLAOs of the less developed state.
4. Senior Level Agricultural Officers (SLAOs) of the developed state versus SLAOs of the less developed state.

The results of discriminant function analysis are presented below.

Combined Employees of the Developed State versus Combined Employees of the Less Developed State

Table 10.1 presents the results of discriminant analysis with regard to combined employees of the developed state versus less developed state. The values of D^2 and 'F' ratio calculated were 6.733 and 17.060 respectively. The value of D^2 showed a significant discriminatory power at 0.01 level of probability. This implied that the variables selected for the discriminant function were useful in discriminating the employees of the developed state as well as the less developed state. The relative importance of the characteristics based on their power to discriminate between the groups is shown by the percentage contribution of each character to the total distance measured.

TABLE 10.1: Combined employees of the developed state versus less developed state. Percentage Contribution of selected variables to the total distance measured

Sl. No.	Variables	Coefficient (λK)	Mean differencc (dK)	Coefficient × Mean difference λK × dK	Percentage
1	2	3	4	5	6
	Work group functioning				
1.	Group clarity	0.180	1.307 (4.49**)	0.235	3.49
2.	Group process	–0.067	0.007 (0.01)	–0.001	–0.01
3.	Group cohesiveness	0.279	1.564 (6.07**)	0.437	6.49
4.	Group homogeneity	0.153	1.213 (4.13**)	0.186	2.76
	Supervisory functioning				
5.	Work facilitation-subordinate relations	–0.012	3.407 (5.39**)	–0.041	–0.61
6.	Work facilitation-goal setting	0.126	0.727 (2.95**)	0.092	1.37
7.	Work facilitation-problem solving	–0.025	0.140 (0.30)	–0.003	0.04
8.	Decision decentralisation	–0.234	–0.593 (2.25*)	0.139	2.06
9.	Production orientation	0.116	2.740 (6.74**)	0.318	4.72
10.	Control of work	0.122	4.360 (7.58**)	0.532	7.90

(*Contd.*)

TABLE 10.1 *(Contd.)*

1	2	3	4	5	6
11.	Competence	–0.046	0.507 (2.87**)	–0.023	0.34
12.	Participation	0.061	1.653 (5.67**)	0.100	1.49
Time use					
13.	Routine	0.105	0.687 (2.38*)	0.072	1.07
14.	Structure	0.476	1.760 (8.26**)	0.838	12.45
15.	Direction	0.229	1.440 (5.33**)	0.330	4.90
16.	Engagement	0.179	1.967 (5.51**)	0.352	5.23
Pay administration					
17.	Pay administration satisfaction	0.171	1.887 (5.99**)	0.323	4.80
18.	Pay satisfaction	–0.142	0.773 (2.54*)	–0.110	–1.63
19.	Personal equity	0.140	0.907 (2.75**)	0.127	1.89
20.	External equity	0.187	1.440 (4.52**)	0.269	4.00
21.	Performance pay contingency	–0.018	0.953 (3.21**)	–0.013	–1.52
22.	Creativity climate	0.193	2.760 (7.05**)	0.567	8.42
Adaptability-flexibility					
23.	Adaptability-anticipating problems	0.678	0.593 (6.52**)	0.402	5.97
24.	Awareness of potential technology	0.586	0.873 (7.36**)	0.512	7.60
25.	Adaptability-quickness in adopting new technology	1.044	0.960 (8.41**)	1.002	14.88
26.	Flexibility-coping with emergencies	0.449	0.400 (3.62*)	0.180	2.67
	Total			6.733	100.00

D square value = 6.733, F value = 17.060
't values are given in parentheses
* Significant at 0.05 level of probability.
** Significant at 0.01 level of probability

The results indicated that the most important factors which discriminated both the groups of employees were adaptability-quickness in adopting new technology (14.88%), time use-structure (12.45%), creativity climate (8.42%), supervisory functioning-control of work (7.90%), adaptability awareness about potential technology (7.60%), group cohesiveness (6.49%), adaptability-anticipating problems (5.97%), time use-engagement (5.23%) and time use-direction (4.90%). The employees of the developed state were characterised as having relatively higher score on the above dimensions, while the employees of the less developed state had lower score. These nine variables alone contributed about 94 per cent to the total distance measured.

It is very interesting to see that the characteristics like higher organisational adaptability, time use and supervisory functioning were most important variables in discriminating the employees of the developed state from the less developed state. The organisational adaptability as perceived by the employees had contributed 28.45 per cent (quickness in adopting new technology 14.88%, awareness about potential technology 7.60% and anticipating problems 5.97%) to the total distance measured. As reported earlier the employees of the department of agriculture in the developed state perceived their department as more adaptable and quick in the introduction of new technology to the farmers. The new technologies varied from introduction of latest improved varieties to the post-harvest technology. Further one could see many new innovative programmes of the department of agriculture in the developed state. For example it has made suitable modifications in the functioning of Training and Visit System (T&V) of extension to increase its effectiveness. It has introduced three new projects as well as five approaches to be followed at village level. This is known as '3 project and 5 steps approach'. The department of agriculture of the developed state felt that T&V extension system has neglected the importance of demonstrations to convince farmers as well as extension workers but relied heavily on information transfer through training sessions or contacts. Further it was felt that more importance was given to interpersonal contact neglecting the importance of mass contact. In order to overcome these difficulties the following three projects were included in the T&V System of extension:

1. Technology development and Demonstration Centres.
2. Need based plant protection programmes to demonstrate the cost effectiveness of Integrated Pest Management.
3. Mass contact programme for diffusion and adoption advancement.

For the implementation of the above projects, the following five steps or approaches were to be followed by the village extension workers and agricultural extension offices :

1. Village level planning with active co-operation of farmers.
2. Taking steps for efficient use of available resources.
3. Conducting one hectare or two hectare results demonstrations on various technologies to be introduced.
4. Pre-season village campaign.
5. Demonstration of techniques of integrated pest management.

The example of modification of T&V System of extension suiting to the local needs speaks about the adaptability and innovativeness of the department of agriculture of the developed state. Even in the administration of the programmes the developed state had introduced new techniques like management by objectives and frequent review of work by the method of grading. But in the department of agriculture of the less developed state no such modifications or innovative programmes suiting to the local needs have been initiated. This shows that there is need for the department of agriculture in the less developed state to be more adaptable and innovative. The latest farm technologies have to be introduced more quickly along with innovative approaches in programme formulation and execution. In this regard the functional linkage of the department of agriculture at various levels has to be strengthened with the agricultural universities and the research institutes.

The time use variables together (structure 12.45%, engagement 5.23%, direction 4.90%) had contributed about 23 per cent to the total distance measured. Thus one of the important discriminating factors which discriminated the employees of the developed state from that of the developed state was time use pattern. The employees of the less developed state lacked structuring and organisation of the time as well as direction. They also had problems in initiating work and lacked purpose. This shows the overall weakness of the organisation, poor supervision, and lack of motivation of the employees. This aspect has to be thoroughly studied and the time use aspect of the employees of the less developed state has to be improved.

Other important variables responsible for discriminating between the two groups of employees were creativity climate, supervisory functioning — control of work and group cohesiveness. The employees of the less developed state perceived their department to be relatively poor in creativity climate, as well as in supervision especially in control of work. There is a need to increase the supervisory control of work through advance planning and better monitoring of work among the employees of the less developed state. Further the measures have to be taken in the department of agriculture of less developed state to increase its creativity climate, as well group cohesiveness.

Village Level Workers of the Developed State versus VLWs of the Less Developed State

Table 10.2 presents the results of discriminant analysis of VLWs of developed state versus less developed state. The values of D^2 and 'F' ratio calculated were 20.056 and 13.411 respectively. The value of D^2 showed a significant discriminant power at 0.01 level of probability. This implied that the variables selected for the discriminant function were useful in discriminating the VLWs of department of agriculture of the developed as well as the less developed state. The relative importance of the characteristics based on their power to discriminate between the groups is shown by the percentage contribution of each character to the total distance measured.

TABLE 10.2: VLWs of the developed state versus less developed state. Percentage contribution of selected variables to the total distance measured

Sl. No.	Variables	Coefficient (λK)	Mean difference (dK)	Coefficient × Mean difference $\lambda K \times dK$	Percentage
1	2	3	4	5	6
Work group functioning					
1.	Group clarity	0.417	2.960 (4.61**)	1.234	6.15
2.	Group process	–0.214	–0.820 (-0.95)	0.175	0.87
3.	Group cohesiveness	1.096	1.060 (2.21*)	1.162	5.79
4.	Group homogeneity	–0.709	0.280 (0.54)	–0.199	–0.99
Supervisory functioning					
5.	Work facilitation-subordinate relations	0.118	1.540 (1.231)	0.181	0.90

(*Contd.*)

TABLE 10.2 *(Contd.)*

1	2	3	4	5	6
6.	Work facilitation-goal setting	–0.166	0.200 (0.413)	–0.033	–0.17
7.	Work facilitation-problem solving	0.677	0.120 (2.327*)	0.758	3.78
8.	Decision decentralisation	–0.465	–1.400 (-2.251*)	0.651	3.25
9.	Production orientation	–0.303	1.680 (2.261*)	–0.509	–2.54
10.	Control of work	–0.232	1.060 (0.93)	–0.246	–1.23
11.	Competence	0.490	0.600 (1.45)	0.294	1.47
12.	Participation	0.622	2.460 (4.32**)	1.629	8.12
Time use					
13.	Routine	0.504	1.360 (2.76**)	0.685	3.42
14.	Structure	1.721	2.260 (5.88**)	3.889	19.39
15.	Direction	0.475	2.340 (4.65**)	1.111	5.54
16.	Engagement	0.136	2.420 (4.01**)	0.329	1.64
Pay administration					
17.	Pay administration satisfaction	0.395	3.060 (3.84**)	1.209	6.03
18.	Pay satisfaction	–0.968	0.520 (0.11)	–0.503	–2.51
19.	Personal equity	0.240	0.220 (0.40)	0.053	0.26
20.	External equity	0.477	0.360 (0.66)	0.172	0.86
21.	Performance pay contingency	–0.142	1.800 (3.41)	–0.255	–1.27
22.	Creativity climate	–0.204	2.440 (3.48**)	–0.497	–2.48
Adaptability-flexibility					
23.	Adaptability-anticipating problems	3.097	0.960 (5.07**)	2.973	14.82

(Contd.)

TABLE 10.2 *(Contd.)*

1	2	3	4	5	6
24.	Awareness of potential technology	1.749	1.400 (7.26**)	2.449	12.21
25.	Adaptability-quickness in adopting new technology	2.037	1.380 (6.80**)	2.811	14.02
26.	Flexibility-coping with emergencies	0.667	0.800 (4.06**)	0.533	2.66
	Total			20.056	100.00

D square value = 20.056, F value = 13.411
't values are given in parentheses
* Significant at 0.05 level of probability.
** Significant at 0.01 level of probability

The results indicated that the most important variables which discriminated both the groups of employees were time use-structure (19.39%), adaptability-anticipating problems (14.82%), adaptability-quickness in introduction of new technology (14.02%), adaptability-awareness about new technology (12.21%), supervisory functioning-participation (8.12%), group goal clarity (6.15%), pay administration satisfaction (6.02%), group cohesiveness (5.79%) and time use-direction (5.54%).

The above nine variables contributed as much as 92.07 per cent to the total distance measured. The VLWs of the developed state were characterised as having relatively higher score in the above dimensions while the VLWs of the less developed state had lower score.

The organisational adaptability (anticipating problems, quickness in the introduction of new technology and awareness about potential technology) as perceived by the VLWs contributed 41.05 per cent to the total distance measured. As reported in the case of combined employees of the department of agriculture, the adaptability of organisation has come out to be the most important factor which discriminated the VLWs of the developed state from the VLWs of the less developed state. This again re-affirms the need of the department of agriculture in the less developed state to be adaptable to the changing environment and be more innovative in its functioning.

The second important group of variables which discriminated the two groups of VLWs was time use (structure and direction) which contributed about 25 per cent to the total distance measured. This clearly points out the need of improving the time structuring and direction of the grass root functionaries (VLWs) of the less developed state. They have problems of organisation of their time and activities and lack direction. This reflect that the existing structure and overall functioning of the department of agriculture of the less developed state has failed to provide effective time use pattern for the grass root functionaries.

Further the results indicated that other variables which discriminated the VLWs of the developed state from their counterparts in the less developed state were participatory supervisory functioning, work group goal clarity and cohesiveness. This implies the need of the department of agriculture of the less developed state to go in for more participatory supervision as well as taking measures to improve the work group cohesiveness and goal clarity.

Junior Level Agricultural Officers of the Developed State versus JLAOs of the Less Developed State

Table 10.3 presents the results of discriminant analysis of JLAOs of developed state versus JLAOs of less developed state. Junior Level Agricultural Officers belong to the cadre of Assistant Director of Agriculture or Agricultural Extension Officer. The values of D^2 and 'F' ratio calculated were 14.116 and 9.178 respectively. The value of D^2 showed a significant discriminating power at 0.01 level of probability. This implied that the variables selected for the discriminant function were useful in discriminating the JLAOs of department of agriculture of the developed state from that of the less developed state. The relative importance of the characteristics based on their power to discriminate between the above groups of employees is shown by the percentage contribution of each character to the total distance measured.

The results indicated that the most important variable which discriminated both the groups of employees was supervisory functioning — control of work (22.67%) followed by creativity climate (16.16%), work group homogeneity (16.05%), adaptability-quickness in introduction of new technologies (6.84%), supervisory functioning-production orientation (6.70%) and supervisory functioning-goal setting (4.92%). The JLAOs of the developed state were characterised as having relatively higher score on the above dimensions than that of JLAOs of the less developed state. These six variables alone contributed about 74 per cent to the total distance measured.

TABLE 10.3: Junior Level Agricultural Officers (JLAOs) of the developed state versus less developed state. Percentage contribution of selected variables to the total distance measured

Sl. No.	Variables	Coefficient (λK)	Mean difference (dK)	Coefficient × Mean difference λK × dK	Percentage
1	2	3	4	5	6
Work group functioning					
1.	Group clarity	0.934	0.660 (1.79)	0.616	4.36
2.	Group process	–0.133	–0.180 (-0.15)	0.024	0.17
3.	Group cohesiveness	0.087	1.780 (3.49**)	0.155	1.10
4.	Group homogeneity	0.0944	2.400 (4.83**)	2.266	16.05
Supervisory functioning					
5.	Work facilitation-subordinate relations	0.143	2.860 (2.78**)	0.409	2.90
6.	Work facilitation-goal setting	0.588	1.180 (2.74**)	0.694	4.92
7.	Work facilitation-problem solving	0.434	–0.100 (-0.28)	–0.043	–0.31
8.	Decision decentralisation	–0.523	–1.160 (-2.36**)	0.607	4.30
9.	Production orientation	0.309	3.060 (4.84**)	0.946	6.70
10.	Control of work	0.711	4.500 (5.21**)	3.200	22.67
11.	Competence	–1.54	0.080 (0.39)	–0.123	–0.87
12.	Participation	–0.107	1.400 (3.21**)	–0.150	–1.06
Time use					
13.	Routine	0.366	0.700 (1.38)	0.256	1.81
14.	Structure	0.379	1.420 (3.99**)	0.538	3.81
15.	Direction	0.211	1.420 (3.14**)	0.300	2.13
16.	Engagement	0.352	1.660 (2.67**)	0.584	4.14

(*Contd.*)

TABLE 10.3 *(Contd.)*

1	2	3	4	5	6
Pay administration					
17.	Pay administration satisfaction	0.342	1.000 (1.29)	0.342	2.42
18.	Pay satisfaction	–0.069	1.380 (1.29)	–0.095	–0.67
19.	Personal equity	0.265	0.680 (1.40)	0.180	1.28
20.	External equity	–0.036	2.600 (4.48**)	–0.094	–0.67
21.	Performance pay contingency	0.184	1.040 (2.17*)	0.191	1.35
22.	Creativity climate	0.610	3.740 (5.47**)	2.281	16.16
Adaptability-flexibility					
23.	Adaptability-anticipating problems	0.156	0.260 (1.42*)	0.041	0.29
24.	Awareness of potential technology	–0.961	0.100 (0.48)	–0.096	–0.68
25.	Adaptability-quick ness in adopting new technology	2.413	0.400 (2.07*)	0.965	6.84
26.	Flexibility-coping with emergencies	0.406	0.300 (1.87)	0.122	0.86
	Total			14.116	100.00

D square value = 14.166, F value = 9.178
't' values are given in parentheses
* Significant at 0.05 level of probability
** Significant at 0.01 level of probability

The supervisory functioning as perceived by JLAOs contributed about 34 per cent (control of work 22.67%, production orientation 6.70% and work facilitation-goal setting 4.92%) to the total distance measured. This means that the JLAOs of the developed state had perceived a supervisory style which emphasised greater control of work through advanced planning and efficient handling of administrative work. They also perceived their supervisors as having greater production orientation. This shows that the department of agriculture of the developed state was relatively better in supervision as perceived by JLAOs. This calls for the need of improving the supervisory effectiveness in the department of agriculture of the less developed state, especially in the area of control of work and goal setting.

Another important factor which discriminated the two groups was their perception of the creativity climate in the organisation. The JLAOs of the developed state perceived their department as having relatively better overall creativity climate. This implies that the work environment of the department of agriculture in the developed state was having more provisions for stimulation of original thinking, greater encouragement for creativity, positive feed back on creativity efforts and opportunities for higher creativity, positive feed back on creativity efforts and opportunities for higher training and discontent with status quo. Perhaps the greater adaptability and innovativeness of department of agriculture in the developed state might have contributed to a better creativity climate than that of its counterpart in the less developed state. This calls for taking up necessary measures in the department of agriculture in the less developed state which will encourage more creativity among its employees.

The other important character which discriminated both the groups was work group homogeneity. The JLAOs of the developed state and perceived a higher group homogeneity. This means the JLAOs in the developed state were more homogenous in their skill, ability and educational background. The homogeneity helps in the smooth functioning of the work group. It was observed that in the developed state all JLAOs were directly recruited and a minimum qualification of B.Sc. (Ag.) was fixed. The graduates of agriculture were directly appointed to the cadre of JLAO who heads the agricultural office at Block level. The promotees from the cadre of VLWs were given a scale which was little lower than that of junior level agricultural officer. Thus the recruitment and promotional policies in the department of agriculture of the developed state made JLAOs to be more homogenous in their age, skill, abilities and educational background. In contrast it was observed that the JLAOs of the less developed state came from many streams partly by direct recruitment and party through promotion. The promotees were high in age but less in qualifications. Further at Block level there were two cadres of agricultural Extension Officer, one known as Agricultural Development Officer and another Senior Agricultural Development Officer who heads the Block unit. Thus there is a need to take measures in the recruitment and promotional policies so that the JLAOs of the less developed sate are homogenous with regard to their educational background, technical skill and training.

Senior Level Agricultural Officers (SLAOs) of the Developed State versus SLAOs of the Less Developed State

Table 10.4 presents the results of discriminant analysis of SLAOs of the developed state versus SLAOs of the less developed state. The Senior Level Agricultural Officers are those officers who belong to the cadre of Deputy Director of Agriculture or above. The values of D^2 and F-ratio calculated were 24.362 and 7.456 respectively. The value of D^2 shows a significant discriminating power at 0.01 level of probability. This implies that the variables selected for the discriminant function are useful in discriminating the SLAOs of the developed state from the less developed state. The relative importance of the characters based on their power to discriminate between the two groups is shown by the percentage contribution of each character to the total distance measured.

TABLE 10.4: Senior Level Agricultural Officers (SLAOs) of the developed state versus SLAOs of the less developed state. Percentage contribution of selected variables to the total distance measured

Sl. No.	Variables	Coefficient (λK)	Mean difference (dK)	Coefficient × Mean difference λK × dK	Percentage
1	2	3	4	5	6
Work group functioning					
1.	Group clarity	0.382	0.033 (0.08)	0.013	0.05
2.	Group process	–0.314	–1.866 (-2.32*)	0.586	2.41
3.	Group cohesiveness	1.765	1.333 (3.52**)	2.352	9.65
4.	Group homogeneity	0.952	1.466 (3.16**)	1.396	5.73
Supervisory functioning					
5.	Work facilitation-subordinate relations	0.457	2.866 (3.18**)	1.310	5.38
6.	Work facilitation-goal setting	0.230	0.933 (2.08*)	0.215	0.88
7.	Work facilitation-problem solving	–0.258	1.833 (-0.91)	–0.473	–1.94
8.	Decision decentralisation	0.039	1.733 (2.93**)	0.067	0.28
9.	Production orientation	0.875	5.553 (6.79**)	4.858	19.94
10.	Control of work	0.652	7.233 (6.45**)	4.716	19.36

(Contd.)

TABLE 10.4 *(Contd.)*

1	2	3	4	5	6
11.	Competence	0.613	2.133 (3.39**)	1.307	5.36
12.	Participation	–0.232	0.833 (1.37)	–0.193	–0.79
Time use					
13.	Routine	–0.444	–0.567 (-1.04)	0.252	1.03
14.	Structure	1.706	1.267 (3.23**)	2.161	8.87
15.	Direction	0.657	0.133 (2.78**)	0.087	0.36
16.	Engagement	–0.706	0.133 (1.79)	–0.094	–0.39
Pay administration					
17.	Pay administration satisfaction	0.570	1.000 (1.42)	0.570	2.34
18.	Pay satisfaction	–0.499	–0.867 (-1.22)	0.433	1.78
19.	Personal equity	–0.242	0.333 (0.42)	–0.080	0.33
20.	External equity	–0.306	–0.433 (-0.79)	0.132	0.54
21.	Performance pay contingency	–0.865	–1.567 (-3.16**)	1.355	5.56
22.	Creativity climate	0.588	1.933 (2.07*)	1.137	4.67
Adaptability-flexibility					
23.	Adaptability-anticipating problems	0.305	0.533 (2.49*)	0.162	0.66
24.	Awareness of potential technology	0.562	0.767 (3.60**)	0.500	2.05
25.	Adaptability-quickness in adopting new technology	0.697	2.100 (4.66**)	1.464	5.99
26.	Flexibility-coping with emergencies	0.552	0.233 (0.97)	0.129	0.53
	Total			24.362	100.00

D square value = 24.362, F value = 7.456

't' values are given in parentheses

* Significant at 0.05 level of probability.

** Significant at 0.01 level of probability

The results indicate that the most important factors which discriminated the SLAOs of the developed state from the SLAOs of the less developed state were supervisory functioning-production orientation (19.94%), control of work (19.36%), group cohesiveness (9.65%), time-use structure (8.87%), adaptability-quickness in the introduction of new technology (5.99%), group homogeneity (5.73%), pay administration-performance pay contingency (5.56%), supervisory functioning-subordinate relations (5.38%) and supervisory functioning-competence (5.37%). The SLAOs of the department of agriculture of the developed state were characterised as having relatively higher score in the above dimensions compared to their counterparts in the less developed state who had lower score. The above nine variables contributed about 86 per cent to the total distance measured.

The group of variables related to supervisory functioning (production orientation, control of work, work facilitation-subordinate relation and competence) contributed in all about 50 per cent to the total distance measured. Thus it is clear that the SLAOs of department of agriculture of the developed state perceived that their department was relatively better in supervisory functioning in terms of supervisors being more production oriented, insisting quality work and also efficient in planning and execution of work. They also perceived the supervisors relatively high in competence, monitoring, and maintaining human relationship with the subordinates. This implies that there is a need for improving the supervisory functioning of department of agriculture of the less developed state even at higher level.

Another group of variables which discriminated the SLAOs of the developed state from that of less developed state was work group functioning (group cohesiveness and group homogeneity) which contributed about 15 per cent to the total distance measured. The SLAOs of the developed state perceived that the work group of their department or units was relatively effective in terms of their cohesiveness and homogeneity. This means that the members of their work group did not vary in their skill and training as well as had higher cohesiveness which would help them to work as a team in achieving the goal of the agricultural department. The above findings call for improving the work group functioning of the units of the department of agriculture of the less developed state in terms of increasing the cohesiveness and homogeneity.

The other important factors which discriminated both the group of employees were time-structuring and perception about the adaptability (quickness in introduction of new technology) of the organisation. Like other employees of department of agriculture, SLAO of the developed state had relatively better position with regard to organisation and structuring of their time. They also perceived their department of agriculture to be effective in the introduction of new technology. This implies that there is a need for effective management of time, especially organisation and structuring of time for the SLAOs of the less developed state. Further as it has been reported earlier the department of agriculture of the less developed state has to become more adaptable and quick in the introduction of new technologies.

The above findings of the discriminant analysis clearly point out that the variables selected in the study are useful in discriminating both the groups of employees. This supports the General Hypothesis, GH-16, which states that the employees of the department of agriculture of the developed state and the less developed state can be distinguished based on selected organisational variables.

11

SUMMARY AND CONCLUSIONS

The really tough part of agricultural development is not the fabrication of improved technologies but rather the organisational task of converting them into production accomplishments. Government machineries for agricultural development in the form of Central and Provincial agricultural departments were established in India after 1871. Even from the beginning, the functioning of agricultural departments has been hampered by lack of effective organisation, finance, trained manpower and appropriate extension methodologies. The organisational set up and the performance of these departments have also been varying from state to state. These differences have widened after Independence. Efficient agricultural administration is one of the important factors which contributes to the overall agricultural development of the states. With all the problems of agricultural development some states like Punjab and Haryana have come up well in agriculture than other states. Does the agricultural administration of the developed states significantly differ from that of less developed states? If yes, in what dimensions? The present study was undertaken to answer the above questions. The specific objectives of the study are as follows:

1. To develop agricultural development index for various states.
2. To study the selected dimensions of organisational context, and macro-organisational structural configuration of state department of agriculture in the developed state and the less developed state.
3. To study the personnel management practices like recruitment, promotion and salary of the employees of the department of agriculture in the developed state and the less developed state.
4. To study the selected dimensions of unit structural characteristics, intra-unit communication, workgroup functioning, job technology, supervisory behaviour, pay administration and creativity climate of state department of agriculture in the developed state and the less developed state.

5. To study the overall organisational effectiveness of agricultural department of the developed state and the less developed state.
6. To find out whether the employees of the department of agriculture of the developed state and the less developed state can be distinguished on the basis of selected variables.

Based upon agricultural development index developed, the states of Tamil Nadu and Madhya Pradesh were selected for the study representing agriculturally developed state and less developed state respectively. For the purpose of studying the organisational structure, processes and effectiveness of the field units of the department of agriculture, the Districts of Madurai and Kamaraj in Tamil Nadu, and Bhopal and Sehore in Madhya Pradesh were selected in consultation with the officials of the department. Apart from the headquarters, all the field units of the department of the agriculture in the above Districts were included in the sample of the study. The proportionate stratified random sampling was used. All the selected units of the department of agriculture in each District including its headquarters were regarded as strata. Each stratum was divided into sub-strata based on the professional ranks such as Deputy Director of Agriculture and above, Agricultural Extension Officer or Assistant Director of Agriculture and Village level extension worker. From each stratum required number of respondents were randomly selected according to the total number of employees of the unit. The number of respondents selected for the study from each state was 150.

The data were collected with the help of an interview schedule developed for the study. The scales or instruments selected for the study have been tested for their reliability and validity.

MAJOR FINDINGS OF THE STUDY

Development of a Composite Agricultural Development Index of Various States of India

On the basis of measures of agricultural development all the 21 states were ranked from developed to less developed in agriculture as follows:

Rank	States	Measures of agricultural development
1.	Punjab	0.1897*
2.	Haryana	0.4656
3.	Tamil Nadu	0.5088
4.	Andhra Pradesh	0.5343
5.	Jammu and Kashmir	0.5533
6.	Kerala	0.6202
7.	West Bengal	0.6467
8.	Gujarat	0.6495
9.	Uttar Pradesh	0.6622
10.	Manipur	0.6966
11.	Karnataka	0.7183
12.	Himachal Pradesh	0.7443
13.	Bihar	0.7530
14.	Maharashtra	0.7671
15.	Meghalaya	0.7677
16.	Tripura	0.7746
17.	Assam	0.7794
18.	Nagaland	0.8289
19.	Orissa	0.8338
20.	Rajasthan	0.8658
21.	Madhya Pradesh	0.8683

* Closer the measure of development to '0' the more developed is the region.

Organisational Context

(a) *Original and History of the Department of Agriculture*

The comparative analysis of origin and history of agricultural departments in Tamil Nadu and Madhya Pradesh showed that Tamil Nadu which is the developed state in agriculture had a relatively early and good beginning in agricultural research than that of Madhya Pradesh. The research work in Central Provinces in the beginning was centred around Nagpur which became the part of state of Maharashtra after reorganisation. The entire area of former Bhopal which was a princely state did not receive sufficient attention in agricultural research or extension before Independence. The total research efforts in terms of number of agricultural research stations and number of scientific workers employed were higher in Tamil Nadu than Madhya Pradesh before Independence. With regard to extension and developmental work both the states had almost a similar organisational set up at State, District and Block level. However, the department of agriculture of Madras province was relatively stronger even in extension work in terms of number of extension workers engaged in extension work at District and Block level. The reasons for the above situation are due to the fact that Madras Province had larger area under cultivation with a higher percentage of area under irrigation. Further due to a higher population density the biotic pressure on land was relatively high in Madras Province which could have necessitated the need for increased attention on agricultural research and extension. The Province of Madras was also relatively prone for general development since it was easily accessible through sea-ports.

The analysis of the origin and growth of the departments of agriculture in the developed and the less developed state showed the importance of contextual factors which influenced the growth of the departments.

(b) *Budgeted Expenditure*

The analysis of budgeted expenditure showed that during pre Independence period (i.e. 1937-38) the per capita expenditure of agricultural department was 4.53 paise and 6.50 paise respectively for Madras and Central Provinces. Though the per capita expenditure was relatively higher in Central Province, it was found that expenditure per acre of gross cropped area was higher in Madras Province (6.98 paise) than Central Province (4.45 paise).

During Sixth Five Year Plan period the average annual expenditure of the agricultural department of Tamil Nadu was 4156.95 lakh rupees and that of Madhya Pradesh was 2403.31 lakh rupees. For the same period the per capita expenditure of the department of agriculture was Rs. 42.92 for Tamil Nadu and Rs. 23.03 for Madhya Pradesh. The expenditure per hectare of gross cropped area for the above period was also much higher in Tamil Nadu (Rs. 321.27) than in Madhya Pradesh (Rs. 56.17).

Macro-Organisational Structure

(a) The department of agriculture of Tamil Nadu was found to be highly differentiated into many other departments or Directorates like Agricultural Engineering, Horticulture and Oilseeds. The department of agriculture of Madhya Pradesh was less differentiated though there was a plan to create separate department like Agricultural Engineering.

(b) Though the organisational set up of the department of agriculture was more or less same in both the states it was found that the Divisional level offices which were present in Madhya Pradesh were missing in Tamil Nadu though every District was treated as a Division headed by a Joint Director. As a result, unlike in Madhya Pradesh the Deputy Directors of Agriculture of Tamil Nadu did not have independent offices but were stationed at the office of Joint Director and functioned mainly as Subject Matter Specialists.

(c) The size of the department of agriculture measured in terms of total number of employees was greater in Madhya Pradesh (30,312 employees) than in Tamil Nadu (14.884 employees).

(d) The VLW (engaged in crop husbandry work)–staff ratio worked out to be 0.698 for Tamil Nadu and 0.602 for Madhya Pradesh.

(e) The number of farm families per VLW under T&V system of extension was relatively lower in Tamil Nadu (1016) than that of Madhya Pradesh (1187).

(f) The number of supervisory level in the department of agriculture of Madhya Pradesh was seven and that of Tamil Nadu was six.

(g) The department of agriculture of Madhya Pradesh had 166 technical job titles and 13 major sections while that of Tamil Nadu had only 48 job titles and seven major sections. Thus the level of horizontal differentiation of the department was higher in Madhya Pradesh.

(h) Though the size of the department of agriculture was 2.03 times greater than that of Tamil Nadu the number of geographical or field units (871) was about 13 per cent lower than that of Tamil Nadu (1016). This was due to the fact that the department of agriculture of Tamil Nadu had 1-2 depots for the supply of farm inputs at every Block.

(i) The supervisor-staff ratio of the department of agriculture of Tamil Nadu (0.047) and Madhya Pradesh (0.042) did not significantly differ from each other and was low. However, the percentage of supervisors above the sub-Divisional unit (above Assistant Director level) to the total supervisors was relatively high in Tamil Nadu (28.7%) than that of Madhya Pradesh (16.21%).

(j) The percentage of post filled into the total number of posts sanctioned (under T&V system of extension) was much higher in Tamil Nadu (98.83%) than that of Madhya Pradesh (62.37%).

(k) The ministerial employees-staff ratio was higher in the department of agriculture of Tamil Nadu (0.328) than that of Madhya Pradesh (0.119).

Recruitment, Salary and Promotional Policies

(a) In Tamil Nadu 90 per cent of VLWs were directly recruited by the department of agriculture and the rest 10 per cent was recruited through the promotion of Field Demonstration Officers. In Madhya Pradesh all the VLWs were directly recruited by Junior Public Service Commission Board. The initial monthly salary of the VLW was relatively higher in Madhya Pradesh (Rs. 965) than that of Tamil Nadu (Rs. 766).

(b) In Tamil Nadu all the Agricultural Extension Officers were directly recruited from among the candidates having a qualification of B.Sc.(Ag.). Their starting monthly salary was Rs. 1243 and their pay scale and official status was equivalent to their counterparts in other department like Animal Husbandry, and Public Works. In Madhya Pradesh about 60 per cent of the total posts of Agricultural Extension Officer was recruited directly and the rest were selected through promotion of village level workers. Their starting monthly salary was Rs. 1123.

(c) While Tamil Nadu had no provision for direct recruitment of officers beyond the level of agricultural extension officers such provisions existed in Madhya Pradesh up to the level of Deputy Director of Agriculture.

(d) The time lag for the promotion of VLW was more or less same (18-20 years) in Tamil Nadu and Madhya Pradesh. However, the time lag for the promotion of Block level Agricultural Extension Officer was relatively high in Madhya Pradesh. It would take about 29 years for a fresh agricultural graduate (who joins as Agricultural Development Officer) to become Assistant Director of Agriculture in Madhya Pradesh through channel of promotion, while it took only about 18 years for the above promotion in Tamil Nadu. Further Tamil Nadu had the provision of 'selection grade' for VLWs as well as Block level Agricultural Extension Officers.

The organisational variables which distinguished the employees of agricultural department of the developed state from that of less developed state have been presented below in a tabular form:

Dimensions	Developed State	Less Developed State
Job Design Characteristics		
1. Job expertise	High	Low
2. Job standardisation	High	Low
3. Job authority	High	Low
4. Job accountability	High	Low
5. Expectation of rewards	High	Low
Unit Structure		
1. Unit standardisation	High	Low
2. Unit employee authority	High	Low
Group Functioning		
1. Group clarity	High	Low
2. Group cohesiveness	High	Low
3. Homogeneity	High	Low
Supervisory Functioning		
1. Work Facilitation subordinate relations	High	Low
2. Work facilitation — goal setting	High	Low
3. Decision decentralisation	High	Low
4. Production orientation	High	Low
5. Control of work	High	Low
6. Competence	High	Low
7. Participation	High	Low
Time — Use		
1. Time structure	High	Low
2. Routine	High	Low
3. Direction	High	Low
4. Engagement	High	Low
Job Technology		
1. Routineness of operation	High	Low
2. Job Inputs adequacy	High	Low
3. Predictability of events	High	Low

(Contd.)

TABLE *(Contd.)*

Dimensions	Developed State	Less Developed State
Pay Administration		
1. Pay administration satisfaction	High	Low
2. Pay satisfaction	High	Low
3. Personal equity	High	Low
4. Performance pay contingency	High	Low
Creativity Climate		
1. Availability of constructive feedback	High	Low
2. Degree of stimulation	High	Low
3. Encouragement and reward to creativity	High	Low
4. Opportunities for higher training	High	Low
5. Availability of physical facilities for experimentation	High	Low
Organisational Effectiveness		
1. Perception of organisation performance	High	Low
2. Organisational adaptability flexibility		
(a) Anticipating problems	High	Low
(b) Awareness of potential technology	High	Low
(c) Quickness in adopting new technologies	High	Low
(d) Coping with emergencies	High	Low
3. Organisational involvement	High	Low

Discriminant Analysis

The discriminant analysis was used to find out the variables discriminating between the employees of the department of agriculture in the developed state and the less developed state. The most important variables which discriminated the two groups of employees were as follows : adaptability-quickness in adopting new technology, time use-structure, creativity climate, supervisory functioning-control of work, adaptability-awareness about potential technology, group cohesiveness, adaptability-anticipating problems, time use-engagement and time use-direction. These nine variables alone contributed about 94 per cent to the distance measured. The employees of the developed state were characterised as having relatively higher score on the above dimensions while the employees of thc less developed state had lower score.

The major conclusion of the study is that agriculturally developed and less developed states can not only be differentiated based on physical inputs like irrigation, fertile land and other resources but also on various dimensions of agricultural administration. The empirical evidences of the study clearly indicate that various dimensions of administrative structure, processes and effectiveness of the department of agriculture of the developed state significantly differed from that of less developed state. The over all organisational effectiveness of the department of agriculture of the developed state was significantly higher than that of less developed state. The study also points out that the growth and functioning of the agricultural department depends upon the different organisational contextual factors which may impose varying constraints as well as facilitate the effective functioning. Thus apart from the organisational constraints the reasons for the relatively poor performance of the agricultural department in the less developed state can be attributed to varying constraints imposed by the contextual factors. The agricultural department can be treated as a system, the functioning and its effectiveness is influenced by its environment.

Implications

The findings of the study are pointers towards a number of implications and these have been discussed below:

(1) The findings of the study show that the disparity or imbalances in the agricultural development of various states of India is quite high. The most agriculturally developed state of Punjab stands well ahead of not only less developed states like Madhya Pradesh or Rajasthan but also relatively developed states like Haryana, Tamil Nadu or Andhra Pradesh. Further all the north-eastern states have lower level of agricultural development. This calls for special attention to be paid for improving the agricultural situation in the less developed states. A long term and massive investment efforts with a system approach are required by all the states especially the states at the bottom of agricul-

tural department to catch up with the development of other states.

(2) The findings indicate that the agricultural departments in the developed and the less developed state have undergone much changes by the way of dividing them into a number of departments like Horticulture, Agricultural Engineering and Oil Seeds. While the formation of specialised agencies is inevitable to effectively cater to the needs of the farmers, it is bound to increase the problems of co-ordination among these agencies as well as inconvenience and confusion among the farmers. Therefore, it is essential that a suitable co-ordination mechanism be established between various agencies of agricultural development at various levels — State, Divisional, District, Block and Village.

(3) The per capita as well as per unit gross cropped area expenditure of the department of agriculture of the developed state was much higher than that of less developed state though the latter had relatively higher per capita expenditure before Independence. Thus it is clear that one of the important characteristics which differentiated the department of agriculture of the less developed state from that of developed state is its smaller size of budget. This calls for increased budget allocation for agricultural department in the less developed state along with the identification of suitable programmes.

(4) It was observed in both the states particularly more in the less developed state that the work of the agricultural department was hindered by the late receipt of financial sanction which limited the amount of work to be done in a particular year. The delay also resulted in hasty incurring of bulk of the expenditure budgeted for the year during the last quarter of the year. The results of such expenditure would not bring desirable effects. This implies that the existing procedures of budgeting and incurring expenditure hinder the efficient functioning of the department of agriculture and these procedures need suitable modifications. An exercise on the study of the current procedures and the formulation of improved have to be undertaken by the department of agriculture with the help of experts in this field.

(5) At the field level it was observed that the lack of financial powers of the officers hindered the efficient functioning of the department of agriculture. This calls for the revising the existing rules and regulations governing the financial powers of different level officers according to the need of the development programmes and changes in the prices of various commodities or inputs.

(6) It was found that the department of agriculture of the developed state had retained in part, the function of supply of farm inputs espe-

cially seeds and pesticides through its depots at Block level. However, it was observed that almost the entire responsibility of supply of farm inputs in the less developed state had been transferred to co-operative organisations. The study reveals that the arrangement of department of agriculture of the developed state in the supply of inputs was working better than that of less developed state which faced a lot of problems in co-ordination between the department of agriculture and the co-operative agency. Though the entire responsibility of handling inputs can not be taken up by the department of agriculture the successful experience of developed state suggest that it should handle a part of the supply with a provision of effective monitoring and co-ordinating the supply of inputs through co-operatives.

(7) The study revealed that the percentage of class I and II posts in the total technical posts of the department of agriculture of the less developed state was about 2-3 times lower than that of other departments like Animal Husbandry, Public Health and Irrigation. This shows that percentage of senior level posts is relatively low in the less developed state. There is a need for correcting this imbalance as this will affect the promotional avenues of the employees.

(8) The supervisor-staff ratio of the department of agriculture of the developed state (0.047) and the less developed state (0.042) was very low and did not differ from each other. This implies that the shape of the pyramid of organisational set up of the departments in both the states is flat having less administrative or supervisory intensity. The lower supervisory-staff ratio is also one of the reasons for poor promotional opportunities for field level extension functionaries. The supervisory intensity of both the departments has to be increased so as to have effective supervision and also to increase the promotional opportunities of the field staff.

(9) The ministerial employees-staff ratio of the department of agriculture of the less developed state (0.119) was far lower than that of the developed state (0.328). Though there are no research data available on the desirable ministerial employees-staff ratio of the development department, it was observed that the department of agriculture of the less developed state faced lot of problems in handling the administrative work of the department. This suggest the need for analysing the existing ministerial staff and improving their position as to increase the efficiency of the administrative wing of the department.

(10) The study revealed that the vacancy position of the technical staff was far higher in the less developed state (37.63%) than in the developed state (1.17%) though there were many unemployed agricul-

tural graduates in the less developed state. This situation had created undue work pressure for other employees and there were many incidences where one employee was holding additional charges of more than 2-3 employees. Immediate attention has to be paid to improve this situation by filling up the posts in time. The functioning and efficiency of the recruiting agency has to be improved.

(11) The time lag for promotion of VLWs and Agricultural Extension Officers was quite high in both the states resulting in poor career development opportunities for those engaged in agricultural extension or development work. The promotional opportunities for the field level extension functionaries have to be increased in order to improve their morale and motivation.

(12) The study also revealed unlike in the case of the developed state, the starting monthly salary and the official status of the Block level Agricultural Extension Officer of the less developed state was lower than that of their counterparts in other departments like Animal Husbandry, Irrigation and Public Works. The pay scale and official status of these officers should be made on par with their counterparts in other departments in order to attract talented people for agricultural extension and developmental work.

(13) It was found that promotion of Agricultural Extension Officers was done only on the basis of seniority in the department of agriculture of the developed state though there was provision for lateral entry in the case of less developed state. There is a need to make provision for recruitment of officers at various levels in the department of agriculture of the developed state so as to increase the career opportunities of the officers and to energise the department with young and talented personnel.

(14) The employees of the department of agriculture of the less developed state when compared to their counterparts in the developed state had relatively lower score on job design dimensions like job expertise, job standardisation, job authority, job accountability and expectation of rewards. The job-design of the employees in the department of agriculture of the less developed state has to be improved so as to increase the job standardisation, authority, accountability and rewards.

(15) The study also revealed that the job inputs adequacy of the employees of the less developed state was relatively lower. This calls for adequate provision of needed job inputs to the employees in the less developed state.

(16) The overall organisational effectiveness of the department of

agriculture of the developed state was significantly higher than that of the less developed state. This implies that the organisational performance of the department of agriculture in the less developed state needs improvement and measures have to be taken to improve the performances.

(17) Both the departments of agriculture have been found to have relatively lower performance in the areas like regularity in field visit, quality of work done, minikit programmes and adaptive trials and reputation for work excellence. Special efforts have to be undertaken to improve the performance in the above areas.

(18) The most important variables which discriminated the department of agriculture of the developed state from that of the less developed state were adaptability-(quickness in adopting new technology and anticipating problems), time use (structure, direction and engagement), creativity climate and supervisory functioning (control of work). The department of agriculture of the developed state was found to be more quick in introducing new technologies to the farmers and had a climate which was more favourable for creative ideas. Further the employees of the department were having higher direction, structure and purposes in their work. And the department had close supervision in guiding and controlling the work of the employees. But the department of agriculture of the less developed state was relatively poor in the above areas. Therefore, it is essential that the department increases its innovativeness, time structuring, creativity climate and supervisory effectiveness so as to improve its functioning and effectiveness.

Inklings for Future Research

The present endeavour is one of the few indepth studies into the understanding of structure and functioning of agricultural departments. Although the pioneering work has come out with useful results, further studies in the following areas are needed to fully comprehend the agricultural development administration :

(1) The study has to be replicated in many other states to draw various generalisations about the agricultural administration in the developed and the less developed states.

(2) There is a need to study the factors contributing to the performance of the department of agriculture based on some predictive analysis.

(3) Action research may be undertaken by manipulating the vari-

ables which discriminated between the developed and the less developed state, to find out the effect of such variables on organisational performance.

(4) There is a need to develop a better design of organisation of agricultural department to improve its overall performance.

(5) An indepth study on the administration of specific programmes or projects, may be undertaken.

APPENDIX-I

INTERVIEW SCHEDULE

DR. Y.P. SINGH
HEAD

DIVISION OF AGRICULTURAL EXTENSION
INDIAN AGRICULTURAL RESEARCH INSTITUTE
NEW DELHI-110012

D.O. No. 9749: Date : 29-3-86

Dear Sir/Madam,

You will be delighted to know that you have been selected as one of the *important respondents* for the research project of our Institute. This research is conducted with the kind permission of Director (Agriculture) of your state.

The success of this research project depends upon your kind co-operation, patience in answering the questions and honest responses. I will be personally grateful if you can spare your valuable time and extend your help and co-operation necessary to Sh. K. Vijayaragavan, a Scientist of our Institute, who will be visiting you for this research work.

There is no right or wrong answer to the questions included in the Schedule. We are interested to know what you think and feel about your work and its environment. We assure that your responses will not be shown to anybody. They will be strictly treated as confidential. You need not write your name. Thank you in advance for your co-operation and help.

With regards,

Yours sincerely,

(Y.P. Singh)

General Instruction : You are required to check (√) one of several responses which are numbered. Select the response with which you agree most. Two examples of such are given below:

1. I like playing cricket.

Strongly disagree	(1)
Disagree	(2)
Slightly disagree	(3)
Slightly agree	(4)
Agree	(5√)
Strongly agree	(6)
Undecided	(7)

2. What colour do you like the most ?

Yellow	Blue	Red	White	Brown
(1)	(2√)	(3)	(4)	(5)

Give your response to the following questions :

1. Are you (check one) (1) Male (2) Female
2. What is your age as on April 1986 : ... yrs...... months
3. State your native state ? ....
4. What is your family size ? ....
5. (a) How many years of total service you have with the present employer ?

   .... yrs.
6. How many years of total service do you have with the other employer(s) ?

   .... yrs.
7. What is your educational qualification ? (check one)

   (1) High School (1) 10 yrs (b) 10 + 2 yrs
   (2) PUC
   (3) B.Sc./B.A.
   (4) M.Sc./M.A.
   (5) Ph.D.

8. What is your total salary per month in your job (all inclusive salary)?

Rs. ....

9. What is your income per year from other sources ? (Business, agriculture etc.)

Rs. ....

10. How many hours per month do you spend in some kind of reading or training to increase your skills needed to do your work ?

(1)	(2)	(3)	(4)	(5)
Nil	About 1-3 hrs	About 4-6 hrs	About 7-9 hrs	About 10 hrs or more

11. Did you receive orientation training related to your job when you joined this job ? (1) Yes (2) No. If yes how much time ?

(1)	(2)	(3)	(4)	(5)
Less than a week	About a week	About a month	1-3 months	More than 3 months

12. How many years did you study (academic or vocational education) after high school?

Years of study after high school (10 +2)

0 1 2 3 4 5 6 7 8 9

13. To what extent do you follow a standard work procedure or practice in your job ?

(eg : Standard procedure in field, visits, administrative work, report writing etc. to do your major jobs in the past 3 months ?)

(1) To no extent (2) Little extent
(3) Some extent (4) great extent
(5) Very great extent

14. Do you have written rules and procedure for all of your major jobs (e.g. : Procedures for procurement of Inputs, General administration, field visits etc.)?

(1)	(2)	(3)	(4)	(5)
No	For a few jobs	For a moderate jobs	For a majority of jobs	For all jobs

15. How clear, precise and practical are the above rules and procedure ?

(1)	(2)	(3)	(4)	(5)
To a very great extent	Great extent	Moderate extent	Not at all	Practically they all useless

16. To what extent the above rules, procedures are practical and useful in all your work situation ?

(1)	(2)	(3)	(4)	(5)
0-20%	21-40%	41-60%	61-80%	81-100%

17. How clearly do you know the amount and quality of work expected from you ? (amount, quality and timeliness or work)

(1)	(2)	(3)	(4)	(5)
Not at all clear	Unclear to great extent	Somewhat clear	Quite clear	Very clear

18. Do you have the job description chart which tell about the work you have to do ? How clear is your job chart which tells about the achievements you have to make on your job, based on which you will be evaluated ?

(1) No job chart or evaluation
(2) No mention of achievements to be made
(3) Somewhat clear with achievement to be made
(4) Quite clear

(5) Very clear and states achievements to be made

19. How much authority do you have in making each of the following decisions about your work ?

	(1) None	(2) Little	(3) Some	(4) Quite a bit	(5) Very much
(a) Determining what work I perform from day to day ?	(1)	(2)	(3)	(4)	(5)
(b) Setting quota on how much work I have to complete every day.	(1)	(2)	(3)	(4)	(5)
(c) Establishing the procedures and rules about how in my work is to be done ?	(1)	(2)	(3)	(4)	(5)
(d) Determining how the unusual circumstances in my job is to be handled ?	(1)	(2)	(3)	(4)	(5)

20. How heavy was your work pressure during the past 3 months ?

(1)	(2)	(3)	(4)	(5)
Practically no useful work	Not enough work	Just right amount	Much work	Too much work

21. How many months in a year, you are completely occupied and busy with your work ?

(1)	(2)	(3)	(4)	(5)
All the 12 months	9-11 months	8-9 months	5-7 months	Below 5 months

22. Do you know in advance about your future work load ? How far in advance ?

(1)	(2)	(3)	(4)	(5)
Do not know at all	About a day in advance	About a week	About a month	About few months in advance

23. During the past three months how much control did you have in setting the speed of your work ?

(1)	(2)	(3)	(4)	(5)
No control	Very little control	Some	Quite a bit	Very much

24. How hard is to attain the quantity and quality of target that is expected from you ?

(1) I do not what is expected out of me
(2) Quite easy to attain
(3) Quite hard
(4) Very hard but attainable
(5) Impossible to attain

25. How much does your supervisor hold you personally accountable for the work decisions you make in your job ?

(1)	(2)	(3)	(4)	(5)
Not at all	Very little	Some	Quite a bit	Very much

26. Do you know any performance standards (achievements to be made in your total work) against which your work is evaluated ? How far these are reasonable and just?

(1) I do not know any performance standard
(2) Very unreasonable standard
(3) Somewhat reasonable
(4) Quite reasonable
(5) Very much reasonable

27. How much does your supervisor hold you personally responsible for achieving the performance standards in your job ?

(1)	(2)	(3)	(4)	(5)
Not at all	Very little	Some extent	Quite a bit	Very much

28. Please indicate how much you agree with the following statements about your job? (Give your feelings).

	Strongly agree (1)	Agree (2)	Undecided (3)	Disagree (4)	Strongly disagree (5)
(1) It is very hard in this job for me to care much about whether or not the work gets done right	(1)	(2)	(3)	(4)	(5)
(2) I feel a very high degree at personal responsibility for the work I do on this job	(1)	(2)	(3)	(4)	(5)
(3) I feel I should personally take credit or blame for the results of my work on this job	(1)	(2)	(3)	(4)	(5)
(4) My supervisor often lets me know how well I am doing my job	(1)	(2)	(3)	(4)	(5)
(5) My co-workers on this job jever let me know how well I am doing my job	(1)	(2)	(3)	(4)	(5)

29. To what extent you can find yourself from the work you have done, how well you are doing your job?

(1) Not at all
(2) To a little extent
(3) To a considerable extent
(4) To a large extent
(5) To a very large extent

30. During the past three months, how often did you receive suggestions or ideas from your co-workers on your work?

(1) Not once (2) About once a month (3) About once a week
(4) About every day (5) Several times a day

31. (a) Is your work evaluated every year ? Yes (1) No (2)
    (b) Have you received the standards against which your work is evaluated?
    Yes No
    (c) Has your supervisor often discussed with you about achieving the above standards?
    (1) Never discussed with me
    (2) Generally discussed with me
    (3) Specifically and clearly discussed with me
    (4) Discussed with me in a detailed way

32. How often did your supervisor had the discussion with you and receive suggestion for improvement of your work ?

(1)	(2)	(3)	(4)	(5)
Never	Seldom	Often	A lot of time	Very large number of time

33. If you attain the performance level that is expected of you:

		(1) No chan-ce	(2) Small chan-ce	(3) 50% chan-ce	(4) Quite likely	(5) Not at all
(a)	Will you be recognised for your good work ? (eg) Given a special word of appreciation	(1)	(2)	(3)	(4)	(5)
(b)	Will you be considered for promotion ?	(1)	(2)	(3)	(4)	(5)
(c)	Will you be rewarded moneta-rily ? (or with a certificate)	(1)	(2)	(3)	(4)	(5)
	If you do not attain in the performance level :					
(a)	Will be told to improve the work with a warning	(1)	(2)	(3)	(4)	(5)
(b)	Will you be considered for demotion ?	(1)	(2)	(3)	(4)	(5)

34. *Your unit :* Your unit is the particular section or office in which you work. Considering your unit as a whole answer the following questions.

(1) Give the number of job titles in your Unit (i.e. the number of different job designations).

....

(2) During the past six months, how many of your immediate unit sub-ordinates performed the basic works ?

(1)	(2)	(3)	(4)	(5)
No one performed the same work	Only a few	About the half	Many	All

35. How many of your immediate subordinates are qualified to do one another's job?

(1)	(2)	(3)	(4)	(5)
None	Only a few	About half	Many	All

36. How easy would it be to rotate the jobs of your immediate subordinates so that each could do a good job performing the other's task?

(1) Very difficult, most members would need extensive training
(2) Quite difficult
(3) Somewhat difficult
(4) Quite easy
(5) Very easy no members would need retraining

37. During the past three months, how often did your immediate subordinates rotate their jobs by performing one another's work?

(1)	(2)	(3)	(4)	(5)
Not once	About every month	About every week	About every day	About every hour

38. Does the performance of your unit is evaluated every year?

1. Yes 2. No

Whether targets (in numerical and quantified measures) have been set for your unit?

(1)	(2)	(3)	(4)	(5)
No	Only some subjective targets	Some numerical target	Specific target	Very precise target

39. How clearly have specific performance targets been set for your unit in each area of your work ? (eg : increase in yield, distribution of HYV seeds etc.)

(1)	(2)	(3)	(4)	(5)
No target	Target for few areas	Some target for all areas	Specific target for all areas	Very precise target

40. How precisely do the rules, policies and procedures specify how work activities are to be coordinated in your unit ?

(1)	(2)	(3)	(4)	(5)
No specific rules	Mostly general	Somewhat specific	Quite specific	Very specific

41. How often did unit members violate or ignore the operating rules, policies and procedure during past 3 months ?

(1)	(2)	(3)	(4)	(5)
Not once	Very seldom	About half time	Quite often	All the time

42. What per cent of these operating rules, policies and procedures for your unit as a whole are written out in memos, reports or a manual ?

(1) (2) (3) (4) (5)

0-20%	21-40%	41-60%	61-80%	81-100%

43. How strictly are these rules, policies or procedures enforced in your work ?

(1)	(2)	(3)	(4)	(5)
Not at all enforced	Loosely enforced	Somewhat strictly enforced	Quite strictly enforced	Very strictly enforced

44. How much influence do you think each of the following has over working of your unit ? (official policies, decision regarding programmes, allocation of sources evaluation of work etc.)

		None	Little	Some	Quite a bit	Very much
		1	2	3	4	5
a.	Members of your unit	(1)	(2)	(3)	(4)	(5)
b.	Your unit head	(1)	(2)	(3)	(4)	(5)
c.	District Officer (Supervisor)	(1)	(2)	(3)	(4)	(5)
d.	Zonal Officer (Supervisor)	(1)	(2)	(3)	(4)	(5)
e.	Supervisors in the state (Headquarters)	(1)	(2)	(3)	(4)	(5)
f.	Others	(1)	(2)	(3)	(4)	(5)

45. How much influence do you think each of the following has over the internal operations of your unit (day-to-day work, work allotment) ?

a. Members of your unit	(1)	(2)	(3)	(4)	(5)
b. Your unit head	(1)	(2)	(3)	(4)	(5)
c. Distt. Level Officers	(1)	(2)	(3)	(4)	(5)
d. Zonal Level Officers	(1)	(2)	(3)	(4)	(5)
e. Officers at State Head-quarters	(1)	(2)	(3)	(4)	(5)

46. When target performance goals for your unit are attained or surpassed how often do the following happens ?

	Almost never	Seldom	About half the time	Quite often	Almost and always

	1	2	3	4	5
(a) All people in this Unit are rewarded or recognised as a group for their team achievements	(1)	(2)	(3)	(4)	(5)
(b) Specific individuals in this Unit are rewarded or recognised for their individual achievements	(1)	(2)	(3)	(4)	(5)

47. When target performance goals for your Unit are not achieved how often do the following happen ?

(a) All people in the Unit are reprimanded or told to 'shape up' to improve unit performance	(1)	(2)	(3)	(4)	(5)
(b) Specific individuals and your Unit head are reprimanded or told to 'shape up' to improve the individual performance	(1)	(2)	(3)	(4)	(5)

48. You may have a set of people with whom you work closely as a team (eg : a team of 2 to 5 Agricultural Officers and village level workers). Please give your feelings about the working of your team.

Strongly disagree	Disagree	Slightly disagree	Slightly agree	Agree	Strongly agree	Undecided
(1)	(2)	(3)	(4)	(5)	(6)	(7)

(a) I feel I am really a part of my team	(1)	(2)	(3)	(4)	(5)	(6)	(7)
(b) My team knows what it has to do	(1)	(2)	(3)	(4)	(5)	(6)	(7)
(c) People who offer new ideas in the team are likely to get discouragements	(1)	(2)	(3)	(4)	(5)	(6)	(7)

(d)	My team members vary in their skill and training	(1)	(2)	(3)	(4)	(5)	(6)	(7)
(e)	My co-workers are afraid to express their real feelings	(1)	(2)	(3)	(4)	(5)	(6)	(7)
(f)	My team members are clear about our goal	(1)	(2)	(3)	(4)	(5)	(6)	(7)
(g)	If we are to make a decision, everybody is involved in it	(1)	(2)	(3)	(4)	(5)	(6)	(7)
(h)	My work group or team contains members with widely varying socio-economic backgrounds	(1)	(2)	(3)	(4)	(5)	(6)	(7)
(i)	We tell each other the way we are feeling	(1)	(2)	(3)	(4)	(5)	(6)	(7)
(j)	I look forward to be with my team each day	(1)	(2)	(3)	(4)	(5)	(6)	(7)
(k)	There are feelings among the members of my team which tend to separate the group	(1)	(2)	(3)	(4)	(5)	(6)	(7)
(l)	In my group every one's opinion is respected	(1)	(2)	(3)	(4)	(5)	(6)	(7)
(m)	There is constant back-biting among members	(1)	(2)	(3)	(4)	(5)	(6)	(7)

49. YOUR SUPERVISOR

Your supervisor is one to whom you report for duty directly. Please give your opinion about your supervisor.

(a) My supervisor

1. Keeps informed about the way our group thinks and feels about	(1)	(2)	(3)	(4)	(5)	(6)	(7)
2. helps sub-ordinates to develop skills	(1)	(2)	(3)	(4)	(5)	(6)	(7)

3. considers that each sub-ordinate is important and gives respect	(1)	(2)	(3)	(4)	(5)	(6)	(7)
4. makes important decision without consulting me	(1)	(2)	(3)	(4)	(5)	(6)	(7)
5. biased on the basis of six	(1)	(2)	(3)	(4)	(5)	(6)	(7)
6. biased on the basis of caste, religion etc.	(1)	(2)	(3)	(4)	(5)	(6)	(7)
7. makes important decisions without asking subordinate's opinions	(1)	(2)	(3)	(4)	(5)	(6)	(7)
8. insist that I work hard	(1)	(2)	(3)	(4)	(5)	(6)	(7)
9. knows thc technical parts of his/her work well	(1)	(2)	(3)	(4)	(5)	(6)	(7)
10. tends to play favourites	(1)	(2)	(3)	(4)	(5)	(6)	(7)
11. is competent	(1)	(2)	(3)	(4)	(5)	(6)	(7)
12. has the respect of subordinates	(1)	(2)	(3)	(4)	(5)	(6)	(7)

(b) My supervisor

1. encourages me to participate in important decisions	(1)	(2)	(3)	(4)	(5)	(6)	(7)
2. plans out work in advance	(1)	(2)	(3)	(4)	(5)	(6)	(7)
3. keeps me informed of all developments	(1)	(2)	(3)	(4)	(5)	(6)	(7)
4. is fair with me	(1)	(2)	(3)	(4)	(5)	(6)	(7)
5. encourages me to speak when I disagree with his decision	(1)	(2)	(3)	(4)	(5)	(6)	(7)

6. makes sure that I have a clear goal to achieve (1) (2) (3) (4) (5) (6) (7)
7. demands the best out of me (1) (2) (3) (4) (5) (6) (7)
8. handles administrative work well (1) (2) (3) (4) (5) (6) (7)

(c) My supervisor

1. makes it clear how I should do my work (1) (2) (3) (4) (5) (6) (7)
2. demands that I produce quality work (1) (2) (3) (4) (5) (6) (7)
3. helps to solve the problems arising in my work (1) (2) (3) (4) (5) (6) (7)
4. is concerned about my general well being (1) (2) (3) (4) (5) (6) (7)
5. discovers the problems before they get too bad (1) (2) (3) (4) (5) (6) (7)

50. You have to communicate with one another in your unit for the functioning of the organisation. Please give your response to the following questions about communication.

Not once (1)	About 1-3 times a month (2)	About 1-3 times a week (3)	About 1-3 times a day (4)	About every hour (5)

1. To co-ordinate the work of the unit during past three months how often were, written reports or memo sent or received.
   - (a) Between you and unit members (1) (2) (3) (4) (5)
   - (b) Among unit members (1) (2) (3) (4) (5)
   - (c) Between you and people outside your unit (1) (2) (3) (4) (5)
   - (d) your supervisor (1) (2) (3) (4) (5)

2. How frequently did you conduct regularly scheduled staff meetings with your immediate subordinates during the past 3 months.

	Once/ week	2-3 month	once in 2 mon-th	once in 3 month	Not once
3. During the past three months how frequently were you involved in unscheduled meetings to solve specific work problems?	(1)	(2)	(3)	(4)	(5)
(a) With two or more of your subordinates	(1)	(2)	(3)	(4)	(5)
(b) With two or more people from outside of your unit?	(1)	(2)	(3)	(4)	(5)
(c) Your supervisor	(1)	(2)	(3)	(4)	(5)
4. During the past 3 months, how often did work related discussions (face to face or by telephone) occur on a one to one basis?	(1)	(2)	(3)	(4)	(5)

Not once	About 1-3 times a month	About 1-3 time a week	About 1-3 times a day	About every hour
(1)	(2)	(3)	(4)	(5)

(a) Between you and your unit members	(1)	(2)	(3)	(4)	(5)
(b) Among unit members	(1)	(2)	(3)	(4)	(5)
(c) Between you and people outside your unit	(1	(2)	(3)	(4)	(5)

51. You have everyday atleast seven hours of work. What do you do with the time? Please give your frank responses to the following questions.

Al-ways	Often	Some times	Rarely	Never
1	2	3	4	5

1. Do you have a rough plan about your activities for tomorrow? (1) (2) (3) (4) (5)
2. Do you have often find that time just seems to slip away? (1) (2) (3) (4) (5)
3. Do you have a daily routine of work? (1) (2) (3) (4) (5)
4. Do you often feel that you seem to work aimlessly or without any purpose? (1) (2) (3) (4) (5)
5. Do you plan your activities day to day? (1) (2) (3) (4) (5)
6. Do you tend to change your work from one to another activity rather aimlessly? (1) (2) (3) (4) (5)
7. Are you sure what you are going to do next? (1) (2) (3) (4) (5)
8. There are many important work for you every day. Do you feel that you don't want to do the things you have to do? (1) (2) (3) (4) (5)
9. Does your work hamper by unscheduled activities? (1) (2) (3) (4) (5)
10. Do you return home with the satisfaction of doing useful work? (1) (2) (3) (4) (5)
11. Do you feel that the things you have to do just don't seem to matter? (1) (2) (3) (4) (5)
12. Do you have problem in organising the things you have to do? (1) (2) (3) (4) (5)

52. Your pay is an important item in your job. We would like to know how do you feel about your pay system? Please give your opinion to the following statements

Strongly disagree	Disagree	Slightly disagree	Slightly agree	Agree	Strongly agree	Undecided
(1)	(2)	(3)	(4)	(5)	(6)	(7)

1. I am content with the way DA, TA etc. are increased

	every year	(1)	(2)	(3)	(4)	(5)	(6)	(7)
2.	I like the present system of promotion	(1)	(2)	(3)	(4)	(5)	(6)	(7)
3.	Considering my skill, efforts qualification I am satisfied with my pay	(1)	(2)	(3)	(4)	(5)	(6)	(7)
4.	My pay is what it ought to be	(1)	(2)	(3)	(4)	(5)	(6)	(7)
5.	For the same level of job in other government departments people receive more pay	(1)	(2)	(3)	(4)	(5)	(6)	(7)
6.	I am happy with the amount of money I make here	(1)	(2)	(3)	(4)	(5)	(6)	(7)
7.	I don't get sufficient pay in comparison to my skill and qualifications	(1)	(2)	(3)	(4)	(5)	(6)	(7)
8.	Private companies and other corporations pay better than my present employer	(1)	(2)	(3)	(4)	(5)	(6)	(7)
9.	Considering my actual work and effort my pay is fair	(1)	(2)	(3)	(4)	(5)	(6)	(7)
10.	Whether I do work or not I receive my pay regularly without problem	(1)	(2)	(3)	(4)	(5)	(6)	(7)
11.	My pay raise does not depends here on my performance	(1)	(2)	(3)	(4)	(5)	(6)	(7)
12.	I am satisfied with my total pay	(1)	(2)	(3)	(4)	(5)	(6)	(7)

53. Think about your job and your experiences with it. Here are some statements about you and your job. How much do you agree or disagree with each?

1.	All in all I am satisfied with my job and the environments in which I work	(1)	(2)	(3)	(4)	(5)	(6)	(7)
2.	In general I don't like working here	(1)	(2)	(3)	(4)	(5)	(6)	(7)
3.	In general I like working							

	here and the people here	(1)	(2)	(3)	(4)	(5)	(6)	(7)
4.	What happens to the organisation is really important to me	(1)	(2)	(3)	(4)	(5)	(6)	(7)
5.	I don't care about what happens here as long as I get my pay	(1)	(2)	(3)	(4)	(5)	(6)	(7)
6.	The developmental programmes of the department does not mean anything to me personally	(1)	(2)	(3)	(4)	(5)	(6)	(7)
7.	I am least bothered about success or failure of the programmes of the department	(1)	(2)	(3)	(4)	(5)	(6)	(7)

54. Every man has got some creative talent. Are you able to use this talent to improve the functioning of your organisation? Give your responses to the following statements

Strongly agree	Agree	Slightly agree	Disagree	Strongly disagree
(1)	(2)	(3)	(4)	(5)

1.	I will be immediately criticised or punished for doing anything novel here	(1)	(2)	(3)	(4)	(5)
2.	People here give me constructive suggestions for any innovative effort of mine	(1)	(2)	(3)	(4)	(5)
3.	I have no opportunity here to learn new techniques relevant to my work here	(1)	(2)	(3)	(4)	(5)
4.	Here there are ample freedom to hole beliefs or positions different from the official line (management)	(1)	(2)	(3)	(4)	(5)
5.	No body bothers about to know about the works in which I am more interested	(1)	(2)	(3)	(4)	(5)
6.	Great respect is given here to					

obedient or dutiful person	(1)	(2)	(3)	(4)	(5)
7. There are no physical or financial facilities to pursue any work related interest or to test new ideas	(1)	(2)	(3)	(4)	(5)
8. My boss expects me to conform to the office norms and procedures	(1)	(2)	(3)	(4)	(5)

55. How fast is your organisation ready to change and accepted new ideas? You will agree that 'change' is necessary in every organisation. Indicate your opinion to the following statements

1. People in my organisation generally find hard to change old methods of doing things	(1)	(2)	(3)	(4)	(5)
2. People in the organisation anticipate problems in advance and make ways to solve them	(1)	(2)	(3)	(4)	(5)
3. The Department of Agriculture is quick enough to receive the latest technology evolved by agricultural university and research institutes for dissemination	(1)	(2)	(3)	(4)	(5)
4. My organisation keeps up with new welcome changes for its overall administration of programmes as well as its staff	(1)	(2)	(3)	(4)	(5)

56. To do your work you may need a lot of information, materials and help. Please answer the following questions which measures the nature of your work and your work technology Environment.

1. To what extent do you need assistance in doing your work from each of the following

Not at all	To a little extent	To a moderate extent	To a large degree	Very great degree
(1)	(2)	(3)	(4)	(5)

(a) Supervisor of your unit (1) (2) (3) (4) (5)
(b) Members of your unit (1) (2) (3) (4) (5)
(c) Employees of other departments/ units (1) (2) (3) (4) (5)

2. To what extent you need information, data, advice, materials (eg, vehicle, supplies of inputs etc.) from each other to do your work?

(a) Supervisor of your unit (1) (2) (3) (4) (5)
(b) Members of your unit (1) (2) (3) (4) (5)
(c) Employees of other departments/units (1) (2) (3) (4) (5)

3. To what extent or degree the following need your assistance in doing their work?
(a) Supervisor of your unit (1) (2) (3) (4) (5)
(b) Members of your unit (1) (2) (3) (4) (5)
(c) Employees of other departments/units (1) (2) (3) (4) (5)

4. Think of all kinds of events that causes your work (eg, going to meet farmers and offering solutions)

(a) To what extent you are able to anticipate and predict the nature of the events (1) (2) (3) (4) (5)
(b) Do you encounter the same kind of problems to be solved in your work? (1) (2) (3) (4) (5)

5. Your job may require the use of one kind of procedure or another (eg. procedure in visiting farm fields or Administrative work). To what extent these procedures are similar from one day to next?

(1) (2) (3) (4) (5)

6. To what extent you do the same kind of work every day? (1) (2) (3) (4) (5)

7. (a) When you are scheduled to start your work, do you have all materials, facilities, information (eg. transport, inputs, literature, etc.) you need? (1) (2) (3) (4) (5)

   (b) If you do not have any of the above materials, is it readily available if you contact appropriate person? (1) (2) (3) (4) (5)

   (c) How far the quality of work is delayed due to non-availability of above things? (1) (2) (3) (4) (5)

57. On the basis of your experience and estimates how good, would you say, is your organisation in the following factors? We are interested to know your opinion.

Outstanding	Very good	Good	Fair	Poor	So So	Very poor
(1)	(2)	(3)	(4)	(5)	(6)	(7)

(a) Reputation for work-excellence among the farmers	(1)	(2)	(3)	(4)	(5)	(6)	(7)
(b) Provision of advisory service to farmers	(1)	(2)	(3)	(4)	(5)	(6)	(7)
(c) Training farmers	(1)	(2)	(3)	(4)	(5)	(6)	(7)
(d) Training of VLWs and agricultural officers by specialists	(1)	(2)	(3)	(4)	(5)	(6)	(7)
(e) Total quantity of work done by the unit or department	(1)	(2)	(3)	(4)	(5)	(6)	(7)
(f) Quality of work done by your unit or department	(1)	(2)	(3)	(4)	(5)	(6)	(7)
(g) Regularity in field visit	(1)	(2)	(3)	(4)	(5)	(6)	(7)
(h) Supply of inputs	(1)	(2)	(3)	(4)	(5)	(6)	(7)
(i) Conducting adaptive trials and minikit programme	(1)	(2)	(3)	(4)	(5)	(6)	(7)
(j) Achieving the target set for the unit	(1)	(2)	(3)	(4)	(5)	(6)	(7)

APPENDIX-II

Distance Matrix

State	1	2	3	4	5	6	7
1. Andhra Pradesh	0.0000	4.7636	6.1228	2.4308	4.0807	4.4964	3.3471
2. Assam	4.7636	0.0000	2.3957	3.6519	5.2862	2.7374	3.4917
3. Bihar	6.1228	2.3957	0.0000	2.6507	4.4731	2.2841	3.0783
4. Gujarat	2.4308	3.6519	2.6507	0.0000	3.4912	3.1676	3.1202
5. Haryana	4.0807	5.2862	4.4731	3.4912	0.0000	4.5169	2.9215
6. Himachal Pradesh	4.4964	2.7374	2.2841	3.1676	4.5169	0.0000	3.0079
7. Jammu & Kashmir	3.3471	3.4917	3.0783	3.1202	2.9215	3.0079	0.0000
8. Karnataka	3.3484	2.9055	1.9136	1.6240	3.9807	2.3056	3.3618
9. Kerala	4.5540	3.4656	4.3325	4.1506	5.0997	4.1849	2.9419
10. Madhya Pradesh	4.5694	3.1594	2.6246	3.4689	5.7386	3.2972	4.9185
11. Maharashtra	3.8275	2.9832	2.2202	2.0005	4.3802	2.6432	3.9487
12. Manipur	4.6520	2.7669	3.0365	4.3156	4.4322	2.5311	2.6968
13. Meghalaya	4.8217	2.2738	2.9615	3.9983	5.5938	2.6694	3.3616
14. Nagaland	6.3261	4.8130	4.4806	5.2094	6.7521	4.6804	4.9379
15. Orissa	4.8985	2.9789	2.0049	3.0804	5.1584	2.5303	4.3444
16. Punjab	7.5628	9.0903	9.0406	7.6714	4.7696	8.5239	6.5214
17. Rajasthan	4.8832	4.2598	3.1335	3.2769	5.5744	3.9398	5.1501
18. Tamil Nadu	5.0465	5.9126	4.8333	3.5545	4.4448	5.1292	3.4178
19. Tripura	4.6492	2.3042	2.8733	3.5678	5.2463	1.3188	3.3447
20. Uttar Pradesh	2.9688	4.3822	2.3673	3.0024	3.6230	3.2210	3.1637
21. West Bengal	3.1309	2.5771	2.2418	2.4996	3.8826	2.4714	1.8543

State	8	9	10	11	12	13	14
1. Andhra Pradesh	3.3484	4.5540	4.5694	3.8275	4.6520	4.8217	6.3261
2. Assam	2.9055	3.4656	3.1594	2.9832	2.7669	2.2738	4.8130
3. Bihar	1.9136	4.3325	2.6246	2.2202	3.0365	2.9615	4.4806
4. Gujarat	1.6240	4.1506	3.4689	2.0005	4.3156	3.9983	5.2094
5. Haryana	3.9807	5.0997	5.7386	4.3802	4.4322	5.5938	6.7521
6. Himachal Pradesh	2.3056	4.1849	3.2972	2.6432	2.5311	2.6694	4.6804
7. Jammu & Kashmir	3.3618	2.9419	4.9185	3.9487	2.6968	3.3615	4.9379
8. Karnataka	0.0000	4.3261	2.1199	0.8410	3.6729	3.3575	4.8885
9. Kerala	4.3261	0.0000	6.0113	4.8194	4.1454	3.6709	5.7418
10. Madhya Pradesh	2.1199	6.0113	0.0000	1.7263	4.4307	3.5930	4.5906
11. Maharashtra	0.8410	4.8194	1.7263	0.0000	3.6674	3.5709	4.9827
12. Manipur	3.6729	4.1454	4.4307	3.6674	0.0000	2.1983	4.4349
13. Meghalaya	3.3575	3.6709	3.5930	3.5709	2.1983	0.0000	3.1843
14. Nagaland	4.8885	5.7418	4.5906	4.9827	4.4349	3.1843	0.0000
15. Orissa	1.7813	5.4990	1.4971	1.6541	3.6927	3.4006	4.8507
16. Punjab	8.1163	7.4408	9.8920	8.6449	8.2265	9.1045	10.0545
17. Rajasthan	2.3611	6.0773	2.0361	2.0022	5.2291	4.4989	5.0015
18. Tamil Nadu	4.5747	4.9469	5.9527	5.2104	5.3042	5.5990	6.2239
19. Tripura	2.8827	3.8750	3.3125	3.0507	2.2806	2.0880	4.5794
20. Uttar Pradesh	2.7216	5.3659	3.9122	3.1868	3.6919	4.2385	5.0687
21. West Bengal	2.1361	2.7846	3.6060	2.1396	2.6700	2.8013	4.8251

State	15	16	17	18	19	20	21
1. Andhra Pradesh	4.8985	7.5628	4.8832	5.0465	4.6492	2.9688	3.1309
2. Assam	2.9789	9.0903	4.2598	5.9126	2.3042	4.3822	2.5771
3. Bihar	2.0049	9.0406	3.1335	4.8333	2.8733	2.3673	2.2418
4. Gujarat	3.0804	7.6714	3.2769	3.5545	3.5678	3.0024	2.4996
5. Haryana	5.1584	4.7696	5.5744	4.4448	5.2463	3.6230	3.8826
6. Himachal Pradesh	2.5303	8.5239	3.9398	5.1292	1.3188	3.2210	2.4714
7. Jammu & Kashmir	4.3444	6.5214	5.1501	3.4178	3.3447	3.1667	1.8543
8. Karnataka	1.7813	8.1163	2.3611	4.5747	2.8827	2.7216	2.1361
9. Kerala	5.4990	7.4408	6.0773	4.9469	3.8750	5.3659	2.7846
10. Madhya Pradesh	1.4971	9.8920	2.0361	5.9527	3.3125	3.9122	3.6060
11. Maharashtra	1.6541	8.6449	2.0022	5.2104	3.0507	3.1868	2.7396
12. Manipur	3.6927	8.2265	5.2291	5.3042	2.2806	3.6916	2.6700
13. Meghalaya	3.4006	9.1045	4.4989	5.5990	2.0880	4.2385	2.8013
14. Nagaland	4.8507	10.0545	5.0015	6.2239	4.5794	5.0687	4.8251
15. Orissa	0.0000	9.4660	2.1283	5.7298	2.8667	3.1877	3.3108
16. Punjab	9.4660	0.0000	9.7982	6.1830	9.0381	7.4279	7.5387
17. Rajasthan	2.1283	9.7982	0.0000	5.7358	4.2050	4.2933	3.9001
18. Tamil Nadu	5.7298	6.1830	5.7358	0.0000	5.7573	3.7332	3.6809
19. Tripura	2.8667	9.0381	4.2050	5.7573	0.0000	4.3574	2.5797
20. Uttar Pradesh	3.1877	7.4279	4.2933	3.7332	4.3574	0.0000	2.9185
21. West Bengal	3.3108	7.5387	3.9001	3.6809	2.5797	2.9185	0.0000

BIBLIOGRAPHY

Adams, J.S. (1965). "Injustice in social exchange", in L. Berkowitz (ed.), *Advances in Experimental Social Psychology* (Vol. 2), New York : Academic Press, pp. 267-299.

Agarwal, R.D. (1973). *Dynamics of Personnel Management in India,* Bombay : Tata-McGraw Hill Publishing Co. Ltd.

Alicia, L.P. (1979). "Learning and creativity", *Journal of Creative Behaviour,* 13(2) : pp. 127-138.

Argyris, Chris (1957). *Personality and Organization,* New York : Harper and Row.

Argyris, Chris (1964). *Integrating the Individual and the Organization,* New York : Wiley.

Ashe, M.R.B. (1965). "A study of change measured personality adjustment and release of creativity through freedom of expression in writing", *Dissertation Abstracts,* 26(2) : pp. 860-61.

Balk, W.L. (1975). "Improving government productivity : Some policy perspectives", *State Profers. Pap. Admin. Stu.,* 3 : pp. 3-25. Beverly Hills : Sage.

Barnard, C.I., (1938). *The Functions of the Executives,* Cambridge, Mass : Harvard University Press.

Bass, B. (1960). *Leadership, Psychology and Organizational Behaviour,* New York : Harper and Row Publishers.

Bavelas, A. and Barret, D. (1951). "An experimental approach to organisational communication", *Personnel,* 27 : pp. 366-371.

Beer, S. (1966). *Decision and Control : The Meaning of Organizational Research and Management Cybernetics,* New York : Wiley.

Bell, G.D. (1967). "Determinants of span of control", *American Journal of Sociology,* 73 : pp. 100-109.

Bennet, M.K. (1951). "International disparities in consumption levels", *American Economic Review,* XLI : pp. 632-649.

Bennis, W.G. (1971). "Towards a 'truly' scientific management : The concept of organizational health", In J. Ghorpade (ed.), *Assessment of Organizational Effectiveness,* pp. 116-143, Pacific Pabsades, CA : Goodyear Publishing Company.

Berger, Chris, J., and L.L. Cummings (1979). "Organizational structure, attitudes and behaviours", In Barry, M. Staw (ed.), *Research in Organizational Behaviour,* Volume 1, Greenwich, CT : JAI Press, pp. 169-208.

Birnbaum, M.H. (1979). "Procedures for the detection and correction of salary inequities", In T.R. Pezzullo and B.E. Brittingham (eds.), *Salary Equity : Detecting and Estimating Sex Bias in Salaries among College and University Professors,* Lexington, Mass : Lexington Books.

Birnbaum, M.H. (1981). "Reply to McLaughlin : Proper path models for theoretical partialing", *American Psychologist,* 36 : pp. 1193-1195.

Birnbaum, M.H. (1983). "Perceived equity of salary policies", *Journal of Applied Psychology,* 68(1) : pp. 49-59.

Blau, P.M. and R.H. Schoenherr (1971). *The Structure of Organizations,* New York : Basic Books.

Blau, P.M. and W.R. Scott (1962). *Formal organizations : A Comparative Approach.* Scranton, Chandler Publishing Company, p. 94.

Blood, M.R. and C.L. Hulin (1967). "Alienation, environmental characteristics and worker responses", *Journal of Applied Psychology,* 51 : pp. 281-290.

Bowers, D.G. and S.S. Seashore (1966). "Predicting organisational effectiveness with a four factor theory of leadership", *Administrative Science Quarterly,* 11 : pp. 238-263.

Cammann, C., M. Fichman, G.D. Jenkins Jr., and J.R. Klesh (1983). "Assessing the attitudes and perceptions of organizational members", In *Assessing Organizational Change : A Guide to Methods, Measures and Practices,* (eds.), Seashore, S.E., E.E. Lawler 111, P.H. Mirvis and C. Cammann, New York : John Wiley & Sons.

Campbell, J.P. (1973). *Research into the Nature of Organizational Effectiveness : An Endangered Species ?* Working paper, University of Minnesota.

Campbell, J.P. (1977). "On the nature of organizational effectiveness", In Goodman, P.S., J.M. Pennings and Associates (eds.), *New Perspectives on Organisational Effectiveness,* San Francisco : Jossey-Bass.

Collins, O.F. (1964). *The Enterprising Man.* Bureau of Business and Economic Research, Graduate School of Business Administration, Michigan : Michigan State University.

Coulter, P.B. (1979). "Organizational effectiveness in the public sector. The example of municipal fire protection", *Administrative Science Quarterly,* 24 : pp. 65-81.

Dasgupta, B. (1971). "Socio-economic classification of districts — A statistical approach", *Economic and Political Weekly,* 6(33) : August 14, pp. 1763-1773.

Department of Agriculture (1954). *Memoirs of the Department of Agriculture,* Madras : Government Press.

Department of Statistics (1921). *Agricultural Statistics of India,* Vol. 1 (1919-20), Delhi : Manager of Publications.

Dorsey, J.T. (1957). "A communication model for administration", *Administrative Science Quarterly,* 2 : pp. 307-324.

Draft Seventh Five Year Plan (1985-90) and Annual Plan 1985-86, Vol. 1 & 2, Govt. of Madhya Pradesh, Bhopal : Planning, Economics and Statistics Department.

Duncan, R.B. (1972). "Characteristics of Organizational Environments and Perceived Environmental Uncertainty", *Administrative Science Quarterly,* 17(3) : pp. 313-27.

Duncan, R.B. (1973). "Multiple decision-making structures in adopting to environmental uncertainty : The impact of organizational effectiveness", *Human Relations,* 26 : pp. 273-279.

Emery, F.E., and Eric L. Trist (1960). "Socio-technical system", In Churchman, C.W. and M. Verhulst (eds.), *Management Science, Models and Techniques,* Vol. 2, Pergamon, pp. 82-97.

Even, R.L. (1964). "An experimental study of the comparative effect of selected art experiences on the creative performance and attitudes of academically superior students", *Dissertation Abstracts,* 24(11) : pp. 4470-71.

Evidence Taken in the Central Provinces and Berar, Vol. VI (1927). Royal Commission on Agriculture in India, Calcutta : Central Publication Branch.

Evidence Taken in the Madras Presidency, Vol. III (1927). Royal Commission on Agriculture in India, Calcutta : Central Publication Branch.

Famularo, J.J. (1972). *Handbook of Modern Personnel Administration,* New York : McGraw-Hill Book Company.

Fayol, Henri (1949). *General and Industrial Management,* Translated by Constance Storns, London : Pittman.

Feather, N.T. and M.J. Bond (1983). "Time structure and purposeful activity among employed and unemployed university graduates", *Journal of Occupational Psychology,* 56 : pp. 241-254.

Fisher, R.A. (1958). As reported by Somasundaram, D. (1976). *Diagnostic Study of Small Farmers with Respect to New Agricultural Technology and its Effective Communication for Adoption,* Unpublished Ph.D. Thesis, New Delhi : Indian Agricultural Research Institute.

Freeman, J.H. (1973) "Environment, technology, and the administrative interests of manufacturing organizations", *American Sociological Review,* 38 : pp. 750-763.

Friedlander, F. and H. Pickle (1968). "Components of effectiveness in small organizations", *Administrative Science Quarterly,* 13 : pp. 289-304.

Gable, R.W. and J.F. Springer (1976). *Administering Agricultural Development in Asia : A Comparative Analysis of Four National Programs,* Boulder, Colorado : Westview Press.

Georgopoulos, B.S. (1957). *The Normative Structure of Social System : A Study of Organisational Effectiveness.* University of Michigan : Unpublished doctoral dissertation.

Georgopoulos, B.S. and A.S. Tannenbaum (1957). "A study of organisational effectiveness", *American Sociological Review,* 22 : pp. 534-40.

Georgopoulos, B.S. and F.C. Mann (1962). *The Community General Hospital,* New York : Macmillan.

Ghiselli, E.E., and C.W. Brown (1955). *Personnel and Industrial Psychology,* New York : McGraw-Hill.

Ghorpade, J. (ed.). (1971). *Assessment of Organizational Effectiveness.* Pacific Palisades, CA : Goodyear Publishing Company.

Gore, W.J. (1956). "Decision-making in federal field offices". *Public Administration Review,* 16 : pp. 281-290.

Guilford, J.P. (1963). "Intellectual resources and their values as seen by scientists", in C.W. Taylor and F. Barron (Eds.), *Scientific Creativity : Its Recognition and Development,* New York : Wiley. pp. 101-118.

Guilford, J.P. (1965). *Fundamental Statistics in Psychology and Education,* Tokyo : McGraw Hill.

Gulick, L. and Urwick (eds.) (1937). *Papers on the Science of Administration,* New York : Institute of Public Administration.

Guion, Robert M. (1965). *Personnel Testing.* New York : McGraw-Hill.

Hackman, J.R. and C.G. Morris (1975). "Group tasks, group interaction process and group performance effectiveness", in *Advances in*

Experimental Social Psychology (Vol. 7), Berkowitz, L. (ed.), New York : Academy Press.

Hackman, J. Richard, and Edward E. Lawler (1971). "Employee reactions to job characteristics", *Journal of Applied Psychology,* Monograph, 55 : pp. 259-86.

Hackman, J. Richard, and Greg, R. Oldham (1975). "Development of the job diagnostic survey", *Journal of Applied Psychology,* 60(2) : pp. 159-70.

Hage, H. (1965). "An axiomatic theory of organizations". *Administrative Science Quarterly,* 10 : pp. 289-320.

Hage, J. and M. Aiken (1967). "Relationship of centralization to other structural properties", *Administrative Science Quarterly,* 12 : pp. 72-92.

Hage, J. and M. Aiken (1969). "Routine technology, social structure, and organization goals", *Administrative Science Quarterly,* 14 : pp. 366-76.

Hall, Richard H. (1962). "Intraorganizational structural variation : Application of the bureaucratic model", *Administrative Science Quarterly,* 7 : pp. 295-308.

Hall, Richard H., J. Eugene Hass, and Norman J. Johnson (1967). "Organizational size, complexity, and formalization", *American Sociological Review,* 32(6) : 903-911.

Hamner, W.C. (1974). Reinforcement theory and contingency management in organizational settings", in *Organizational Behaviour and Management : A Contingency Approach,* Tosi, H.L. and W.C. Hamner (eds.), Chicago : St. Clair Press.

Haplin, A.W. and B.J. Winer (1957). "A factorial study of the leader behaviour descriptions", In *Leader Behaviour : It's Description and Measurement.* Stogdill, R.M. and A.E. Coons (eds.), Monograph 88, Columbus : Bureau of Business Research, Ohio State University, pp. 39-51.

Harvey, E. (1968). "Technology and the structure of organisations". *American Sociological Review,* 33 : pp. 247-259.

Hatry, H.P. (1972). "Issues in productivity measurement for local governments", *Public Administration Review,* 32(6) : pp. 776-84.

Hatry, H.P. (1978). "The status of productivity measurement in the public sector", *Public Administration Review,* 38(1) : pp. 28-34.

Hatry, H.P., R.E. Winnie and D.M. Fisk (1973). *Practical Program Evaluation for State and Local Government Officials.* Washington DC : Urb. Inst.

Hayes, F. (1978). "City and country productivity programs", *Public Administration Review,* 38(1) : pp. 15-19.

Hickson, D.J., D.S. Pugh, and D.C. Pheysey (1969). "Operations technology and organizational structure : An empirical reappraisal", *Administrative Science Quarterly,* 14 : pp. 378-397.

Hickson, D.J., C.R. Hinings, C.A. Lee, R.E. Schneck, and J.M. Pennings (1991). "Strategic contingencies theory of intraorganizational power", *Administrative Science Quarterly,* 16 : pp. 216-229.

Hovland, C.I., I.L. Janis and H.H. Kelley (1953). *Communication and Persuasion,* New Haven, Conn. : Yale University.

Hovland, C.I. *et al.* (eds.). (1957). *The Order of Presentation in Persuasion. Vol. 1.* Yale studies in attitude and communication, New Haven, Conn. : Yale University.

Hulin, C.L. and M.R. Blood (1968). "Job enlargement, individual differences, and worker responses", *Psychological Bulletin,* 69 : pp. 41-55.

Huxley, A. (1963). "Achieving a perspective on the technological order", *Technology and Culture,* 3 : pp. 636-642.

Imperical Council of Agricultural Research (1941). *Agriculture and Animal Husbandry in India (1938-39),* Delhi : Manager of Publications.

Indik, B.P., B.S. Georgopoulos and S.E. Seashore (1961). "Superior — subordinate relationship and performance", *Personnel Psychology,* 14.

Inkson, J.H.K., D.J. Hickson and D.S. Pugh (1970). "Organisational context and structure : An abbreviated replication", *Administrative Science Quarterly,* 15 : pp. 318-329.

Israeli, N. (1946). "Studies in occupational analysis. II. Originality", *Journal of Psychology,* 22 : pp. 77-87.

Jahoda, M. (1979). The Impact of Unemployment in the 1930's and the 1970's, *Bulletin of the British Psychological Society,* 32 : pp. 309-314.

Jenkins, G.D. (1977). The Impact of Employee Participation in the Development and Implementation of a Base Bay and a Performance Bonus Plan, Unpublished doctoral dissertation, University of Michigan.

Jorgenson, D.O., M.D. Dunnette and R.D. Pritchard (1973). "Effects of the manipulation of a performance — reward contingency on behaviour in a simulated work setting", *Journal of Applied Psychology,* 57 : pp. 271-280.

Kahn, R.L (1956). "The prediction of productivity", *Journal of Social Issues,* 12(2) : pp. 41-49.

Kahn, R.L. (1960). "Productivity and job satisfaction", *Personnel Psychology,* 13 : pp. 275-287.

Kanter, R.M., and D. Brinkerhoff (1981). "Organizational performance : Recent developments in measurements", *Annual Review of Sociology,* 7 : pp. 321-49.

Karoles, V. (1964). "Economic correlates of artistic creativity", *American Journal of Sociology,* 70 : pp. 332-341.

Katz, D., B.A. Gutek, R.L. Kahn and E. Barton (1975). *Bureaucratic Encounters : A Pilot Study in the Evaluation of Government Services,* Ann. Arbov, Mich, Univ. Michigan, Inst. Soc. Res.

Katz, D., N. Macoby, G. Gurin, and L.G. Floor (1951). *Productivity, Supervision and Morale Among Rail Road Workers,* Ann. Arbov, Michigan : Survey Research Centre, University of Michigan.

Katz, D., N. Macoby, and N.C. Morse (1950). *Productivity, supervision and morale in an office situation,* Ann. Arbor, Michigan : Survey Research Centre, University of Michigan.

Katz, D., and R.L. Kahn (1951). "Human organization and worker motivation", In *Industrial Productivity.* Tripp, R. (ed.), Madison, Wis. Industrial Relations Research Association.

Katz, D. and R.L. Kahn (1966). *The Social Psychology of Organizations,* New York : Wiley.

Khandwalla, P.N. (1977). *The Design of Organizations,* New York : Harcourt, Brace, Jovanovich.

Khandwalla, P.N. (1984). *Fourth eye-excellence through creativity,* Allahabad : A.H. Wheeler & Co. Private Limited.

Lawler, E.F. III. (1971). *Pay and Organisational Effectiveness : A Psychological View,* New York : McGraw-Hill.

Lewin, K. (1958). "Group decision and social change". In Macoby, E.E., T.M. New Comb and E.M. Hartley (eds.), *Reading in Social Psychology.* New York : Holt, Rinehart and Winston, pp.197-211.

Likert, R. (1961). *New Patterns of Management*, Tokyo : McGraw-Hill Kogakusha Ltd.

Likert, R. and Kahn, R.L. (1956). "Planning for effective leadership", In *Planning and Training for Effective Leadership,* Seashore, S. (ed.), Ann. Arbor, Michigan : Foundation for Research on Human Behaviour, pp. 1-7.

Lipsksy, M. (1978). "The assault on human services : Street level bureaucrats, accountability and fiscal crisis", In *Accountability in Urban Society : Public Agencies under Fire. Urban Affairs Annual*

Reviews, S. Green, R.D. Hedlund and J.L. Gibson (eds.), 15 : 15-39, Beverly Hills, California : Sage.

Lipsksy, M. (1980). *Street Level Bureaucracy : Dilemmas of the Individual in the Public Services,* NY : Sage Foundation.

Lucey, P.J. (1978). "Wisconsin's progress with productivity improvements", *Public Administration Review,* 38(1) : pp. 9-12.

Lynch, B.P. (1974). "An empirical assessment of Perrow's technology construct", *Administrative Science Quarterly*, 19 : pp. 338-356.

Macdonald, D. and R.V. Farance (1970). *Approaches to the Study of Communication in Organisations : An Example of Applied Theory in a Large Financial Institution,* Paper presented to the Speech Communication Association.

March, J.G. and H.A. Simon (1958). *Organizations,* New York : Wiley.

Mausner and Snyderman (1956) as reported by Van de ven, A.H. and D.L. Ferry (1980). *Measuring and Assessing Organizations,* New York : John Wiley & Sons.

Mann, F.C. (1962). "Towards an understanding of the leadership role in formal organisation", In *Leadership and Productivity,* Dubin, R., G.C. Homans, F.C. Mann and D.C. Miller (eds.), San Francisco, California; Chandler Publishing Company, pp. 68-103.

Mann, F.C. and Dent, J. (1954). *Appraisal of Supervisors and Attitudes of their Employees in an Electric Power Company,* Ann. Arbor, Michigan : Institute for Social Research.

Mann, F.C. and H.J. Baumgartel (1953). *Absences and Employees Attitudes in an Electric Power Company,* Ann. Arbor, Mich. : Institute for Social Research.

Mark, J.A. (1972). "Meanings and measures of productivity", *Public Administration Review,* 32(6) : pp. 747-53.

Miller, J.G. (1965). "Living systems : Basic concepts", *Behavioural Science*, July : pp. 209-211.

Morrissey, E. and D.F. Gillespie (1975). "Technology and the conflict of professionals in bureaucratic organizations", *The Sociological Quarterly,* 16 : pp. 319-322.

Morris, T.D., W.H. Corbett and B.L. Usilanar (1972). "Productivity measures in the federal government", *Public Administration Review,* 32(6) : pp. 753-64.

Morse, J. and J. Lorsch (1970). "Beyond Theory", *Harvard Business Review,* pp. 61-68 (May/June).

Mott, P.M. (1972). *The Characteristics of Effective Organizations,* New York : Harper and Row.

Mukherjee, M. (1980). *Physical Quality of Life Index,* Bombay : Centre for Monitoring Indian Economy.

Nanjappa, M.B. and P. Sudarshan (1981). "A note on inter-district differential in Karnataka development", *J I & W,* 4(2).

Negandhi, A.R. and B.C. Reimann (1973). "Task environment, decentralization and organisational effectiveness", *Human Relations,* 26 : pp. 203-14.

Normann and Rhenman (1975). As reported by Kanter, R.M. and D. Brinkerhoff (1981). "Organizational performance : Recent developments in measurement", *Annual Review of Sociology,* 7 : pp. 321-49.

O'Brien, M.A., L.A. Sibley Jr. and E.M. Ligen (1953). "Developing creativity in children — use of imagination : Theoretical statement", *Union College Studies Character Research,* 1(3) : pp. 17-216.

Oldham, Greg R., J. Richard Hackman, and Jone L. Pearce (1976). "Conditions under which employees respond positively to enriched work", *Journal of Applied Psychology,* 61 : pp. 395-403.

Opsahl, R.L. and M.D. Dunnette (1966). "The role of financial compensation in industrial motivation", *Psychological Bulletin,* 66(2) : pp. 94-118.

Otte, R.W. (1966). "The reactions of elementary school teachers and principals in Indian public school study council school systems to various factors that foster and hamper creativity in teaching", *Dissertation Abstracts,* 25(10) : pp. 5794-95.

Perrow, C. (1967). "A framework for the comparative analysis of organizations", *American Sociological Review,* 32 : pp. 194-208.

Peterson, R.B. (1975). "The interaction of technological process and perceived organizational climate in Norwegian firms", *Academy of Management Journal,* 18 : pp. 288-299.

Pfeffer, J. (1978). *Organizational Design,* Arlington Heights, TII : AHM Publishing.

Pierce, Jon L. (1978). *Employee Affective Responses to Work Unit Structure and Job Design : A Test of an Intervening Variables,* Unpublished paper, Duluth : University of Minnesota, School of Business and Economics.

Porter, L.W. and E.E. Lawler III (1968). *Managerial Attitudes and Performance,* Homewood, IL : Irwin-Dorsey.

Porter, Lyman W., Edward E. Lawler III, and J. Richard Hackman (1975). *Behaviour in Organizations,* New York : McGraw-Hill.

Possien, W.M. (1965). "A comparison of the effects of three teaching methodologies on the development of the problem solving skills of sixth grade children", *Dissertation Abstracts,* 25(7) : p. 4003.

Price, J.L. (1968). *Organisational Effectiveness : An Inventory of Propositions,* Homewood, IL : Irwin.

Price, J.L. (1972). *Handbook of Organizational Measurement,* Lexington, MA Heath.

Pugh, D.S., D.J. Hickson, C.R.Hinings and C. Turner (1969). "The context of organization structure", *Administrative Science Quarterly,* 14 : pp. 91-114.

Rao, S.K. (1973). "A note on measuring economic distances between regions in India", *Economic and Political Weekly,* 8(17) : April 28, pp. 793-800.

Raychaudhuri, M. (1966). *Studies in Artistic Creativity,* Calcutta : Rabindra Bharati, pp. 114-115.

Report of the Agricultural Administration Committee. (1958). New Delhi : Department of Agriculture, Ministry of Food and Agriculture.

Report of the First Indo-American Team on Agricultural Education, Research and Extension (1955). New Delhi. I.C.A.R.

Report of the Royal Commission on Agriculture in India (1928). Bombay : Government Central Press.

Report of the Study Team on Agricultural Administration, Vol. I & II (1967). Administrative Reform Commission Delhi, The Manager of Publications.

Report of India's Food Crisis and Steps to Meet it (1965). New Delhi : Ministry of Food and Agriculture.

Report on the Introduction of Improvement in Indian Agriculture (1909). Calcutta : Manager of Publications.

Roose, J.E. and M.E. Doherty (1978). "A social judgement theoretic approach to sex discrimination in faculty salaries", *Organizational Behaviour and Human Performance,* 22 : pp. 193-215.

Rousseau, D.M. (1983). "Technology in organizations : A constructive review and analytical framework", in *Assessing Organizational Change: A Guide to Methods, Measures and Practices,* Seashore, S.E., E.E. Lawler III, P.H. Mirvis and C. Cammann, (Eds.), New York : John Wiley.

Russel, John (1937). *Report on the Work of the I.C.A.R. in Applying Science to Crop Production,* Delhi : Manager of Publications.

Ryan, R. (1978). "The impact of three years of experience and a new Governor on the State of Washington's productivity program", *Public Administration Review,* 38(1) : pp. 12-15.

Savoca, A.F. (1965). "The effect of reward, race, I.Q., and socio-economic status on creative production of preschool children", *Dissertation Abstracts,* 26(4) : p. 2327.

Schein, E.A. (1970). *Organizational Psychology,* Englewood Cliffs, NJ : Prentice-Hall.

Scott, W.R. (1975). "Organizational structure", *Annual Review of Sociology,* 1 : pp. 1-20.

Seashore, S.E. (1954). *Group Cohesiveness in the Industrial Work Group,* Ann. Arbor, Michigan : Survey Research Centre, University of Michigan.

Seashore, S.E., E.D. Lawler III, P.H. Mirvis and C. Cammann (1983). *Assessing Organizational Change : A Guide to Methods, Measures and Procedures,* New York : John Wiley & Sons.

Spotts, J.B. (1976). "The problem of leadership : A look at some recent findings on behavioural science research", In *Leadership and Social Change.* Lassey, W.R. and R.R. Fernandez (eds.), California : University Associates Inc. p. 56.

Srinivasan, N. (1969). *Agricultural Administration in India.* New Delhi : Indian Institute of Public Administration.

Srivastava, S.C. (1983). "Application of taxonomic method for measuring the level of development of various districts of Uttar Pradesh", In *Regional Dimensions of India's Economic Development — Proceedings of Seminar at Nainital, India,* April 22-24, 1982, Lucknow : State Planning Commission (Planning department).

State Planning Commission (1986). *Seventh Five Year Plan (1985-90).* Tamil Nadu Madras : State Planning Commission.

State Planning Commission, Tamil Nadu (1986). *Draft Annual Plan 1985-86, Vol. I & II,* Madras : State Planning Commission.

Statistical Abstract for British India — 1924-25 to 1933-34 (1936). Delhi : Manager of Publications.

Steers, R.M. (1975). "Problems in the measurement of organizational effectiveness", *Administrative Science Quarterly,* 20 : pp. 546-558.

Stein, M.I. (1962). "Creativity and the scientist", in *The Sociology of Science.* B. Barber and W. Hirsch (eds.). New York : Free Press, pp. 329-343.

Taylor, Frederick W. (1911). *The Principles of Scientific Management,* New York : Harper and Row.

Terborg, J.R. and H.E. Miller (1978). "Motivation, behaviour, and performance : A closer examination of goal setting and monetary incentives", *Journal of Applied Psychology,* 63(1) : pp. 29-39.

Thompson, J.D. (1967). *Organizations in Action,* New York : McGraw-Hill,.

Trist, Eric L. (1970). *A Socio-Technical Critique of Scientific Management,* Paper presented at the Edinburgh Conference on the Impact of Science and Technology, Edinburgh University (May 24-26).

Tyler, W.B. (1973). "Measuring organizational specialization : The concept of role variety", *Administrative Science Quarterly,* 18(3) : pp. 383-92.

Untermeyer, L. (1955). *Makers of the Modern World,* New York : Simon and Schuster.

Van de ven, A.H. (1976). "A framework for organization assessment", *Academy of Management Review,* 1(1) : 64-78.

Van de ven, A.H., D. Emmett and R. Koening Jr. (1974). "Frame works for inter-organizational analysis", *Organization and Administrative Sciences Journal,* 5(1) : pp. 113-29.

Van de ven, A.H. and D.L. Ferry (1980). *Measuring and Assessing Organizations,* New York : John Wiley & Sons.

Voelcker, John Augustus (1897). *Report on the Improvement of Indian Agriculture,* Second Edition, Calcutta : Office of Superintendent of Government Printing.

Webb, R.J. (1974). "Organization effectiveness and the voluntary organization", *Academy of Management Journal,* 17 : pp. 663-677.

Webber, Max (1947). *The Theory of Social and Economic Organizations,* Translated by A.M. Hinderson and T.M. Parsons, New York : The Free Press.

Willits, R.D. (1967). "Company performance and interpersonal relations", *Industrial Management Review,* 7.

Wood, M.T. (1973). "Power relationships and group decision-making in organizations", *Psychological Bulletin,* 79 : p. 280.

Woodward, J. (1958). *Management and Technology,* London : Her Majesty's Stationery Office.

Yuchtman, E. and S.E. Seashore (1967). "A system resource approach to organisational effectiveness", *Am. Social. Rev.,* 32 : pp. 891-903.

Zacharias, C.W.B. (1950). *Madras Agriculture,* Madras : University of Madras.

INDEX